Digital Convergence Phase 2

A Field Guide for Creator-Collaborators

By

Andy Covell

Published by

Stipes Publishing L.L.C.
204 West University Avenue
Champaign, Illinois 61820

Library of Congress Catalogue Number: 2004107946

Digital Convergence* Phase 2: *A Field Guide for Creator-Collaborators

ISBN 1-58874-372-1

Stipes Publishing L.L.C.
202-204 West University Avenue
Champaign, Illinois 61820

Printed in the United States

10 9 8 7 6 5 4 3 2 1

TABLE OF CONTENTS

ACKNOWLEDGMENTS

I wish to thank the many students, colleagues, and friends who helped me in some way to write this book. I also want to offer a special thank-you to my family for their support, and I want to express my sincere gratitude to the following individuals for their gracious contributions of information, advice, and feedback: Cameron Clarke, Kirsten Dennison, Ray Dow, Pradeep Jotwani, Phil Leigh, Dan McClung, Gary McGinnis, Mark Miller, Dave Molta, Mary Peery, Jason Peyser, Todd Rosenberg, Chris Sedore, Tom Streeter, and Steven Woodard.

Part I

A Field Guide

Todd Rosenberg's story captures the impact and opportunities of the digital convergence phenomenon.

Todd used to be a Biz Dev (a.k.a. Business Development) professional in the dot.com sector. As a dot.com carpetbagger, he worked for two years at BarnesandNoble.com and then a year at AtomFilms.com, an Internet outlet for up and coming independent filmmakers. He tried to do deals. He did a lot of lunches. In the summer of 2001, Atom Films (like so many other dot.coms) laid off seventy-five percent of its work force. They closed the New York office where Todd worked.

Todd had some time and he needed some money. A creative guy who had some experience as a cartoonist, he decided to put his creativity to work on the Web. He took the moniker Odd Todd and created a one-person organization called No Nothing Productions. He set up a Web site (www.oddtodd.com) that set him back $9.95 a month, and populated it with humorous audio rants about being laid off.

He decided that animation could strengthen his site, so he bought a copy of Flash, the hugely popular animation software from Macromedia. He spent several weeks learning Flash and putting up Laid Off: A Day in the Life, *a cartoon depicting a day in the life of laid off Odd Todd. He sent the Web address via email to fifty people. While the site and the cartoon were free, Todd included a tip jar using PayPal and the Amazon Honor System payment mechanisms. He also rented a P.O. box so folks could send him tips.*

When I spoke to Todd in January 2002, he had already seen one million unique visitors to his site, and he'd made $7,000 in tips. He had some freelance work, and his Laid Off *animation was to be featured in a movie. He'd been interviewed by radio stations and written up in major newspapers such as* USA Today. *Not long after, he turned up on the* Today *show. He told me this was a life-altering experience. Odd Todd had experienced the dot.com gold rush and the subsequent nose dive. He then experienced the empowering impact of powerful convergence technologies.*

This is a field guide for those of you who are unafraid of technology; are creative and collaborative; and hopefully have a little bit of Odd Todd in you.

This field guide is not *meant only to help you identify components of the emerging digital environment – so you can quickly and visually distinguish between, say, a Palm PDA and a Pocket PC, the way you might use a bird watcher's field guide to distinguish between the various finches. It is more like the old* Audubon Society Field Guide *I have in my home whose purpose is "to enable the reader to explore and enjoy the natural history and ecology of our region." I want you to enjoy, explore, and profit from the new capabilities afforded by emerging digital technologies in this, the age of digital convergence.*

1. INTRODUCTION

I'm on a mission to do what I can to place the power of new technology into the hands of people. To that end, I'm building a body of work that chronicles the technology-based rise of the individual in our economy and society. I fully intend for this to inspire and guide my readers in the pursuit of promising new technologies.

There are millions of people on this earth who regularly use a variety of new technologies, understand some of the high-tech buzzwords, and recognize that something important is happening as digital technologies evolve and converge – something that offers potent individual opportunities.

I have worked in computing at Syracuse University since 1981 and am now director of information technology at the Martin J. Whitman School of Management. In the late 1990s I became convinced that many people could benefit from a conceptual framework that clarified the nature and potential of new technologies, a framework that builds confidence and serves as a foundation for action. I wrote *Digital Convergence*[*] to provide that framework.

Digital Convergence is a technology primer focused on the evolution and convergence of underlying digital computing, communications, and media technologies. It also explains some important application technologies – such as the Web, videoconferencing, and streaming media – that enable individuals to interact, communicate, collaborate, and share information in new ways. I wanted to help people understand the technology and its evolution, stimulating ideas and ambitions in the process – in short, to empower with knowledge and understanding.

---

[*] *Digital Convergence: How the Merging of Computers, Communications, and Multimedia is Transforming Our Lives*, Aegis Publishing, December, 1999.

Since I wrote *Digital Convergence*, which will henceforth be referred to as "the first book," the evolution and convergence of technologies has progressed. We've also begun to see more clearly what convergence enables – and what it does not.

- We have witnessed the rise, and then the demise, of the dot.coms, a phenomenon marked by the failed efforts of venture capitalists and entrepreneurs to seize control of the Internet as a mass medium and mass market.
- The Internet infrastructure has matured, and we're finally to the point where high-speed Internet access for consumers and small businesses – commonly referred to as broadband – is becoming more broadly deployed.
- We have continued to endure the rough edges of new technology – a lack of reliability, security, privacy, and ease of use – that have stubbornly refused to yield.
- We have seen a steady stream of new convergence technologies that empower individuals, workgroups, and entrepreneurs (hopefully a firsthand experience for many of you).

Since the first book we've also seen new rich media technologies (audio, video, animations, etc.) begin to take hold, and we've seen incredible hype for the prospects of wireless mobile Internet access. These developments mark the beginning of a new phase of the digital convergence era.

Like the first book, this book is first and foremost a book about technology. It further explores the technological phenomenon of digital convergence, a conceptual framework that captures the most significant characteristics and realities of the digital revolution, from a technological standpoint. But it's also a book about people.

Why? Because as the revolution in technology unfolds, it is clear that individuals like you and I are the most likely beneficiaries. I firmly believe that we will be the ultimate big winners, even as Microsoft, AOL Time Warner, Cisco, and many other large and small companies secure their share of the economic benefits. When I wrote the first book, I concluded with a section titled "Here's to the Little Guy" which presented the potential of convergence technologies to empower. This book picks up where that left off, since I'm now convinced that the empowering impact of new technology on individuals – how we learn, work, and play – is where the biggest paradigm shifts exist.

This is a book for and about the people who can benefit from these paradigm shifts: individual creators and collaborators who are already leveraging new technologies for personal and professional profit; and passive consumers who may want to step up to the plate to take a swing at the nice fat hanging curveball that digital convergence is serving up.

Digital Convergence

Before we get any further into laying out what the book offers, I need to define digital convergence for those of you who haven't read the first book (see Figure 1). I also need to update this conceptual framework – since we're now moving into the second phase of the digital convergence era.

> **Digital Convergence**: the convergence of computing, digital communications technology, and digital media (technologies and content).

Figure 1

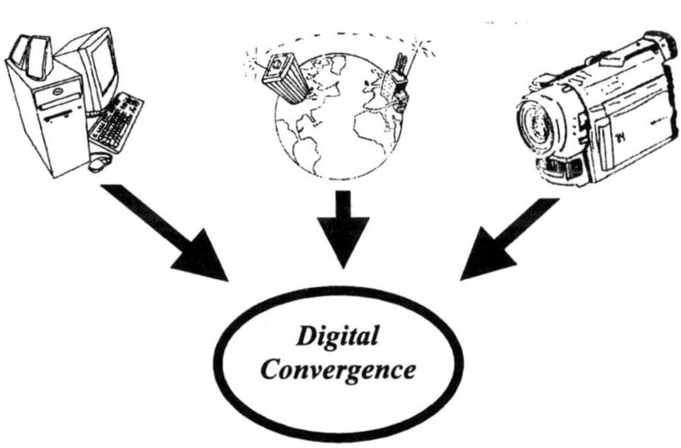

When these three technological realms converge, wonderful new capabilities, resources, and tools are created. The digital convergence era was born in the 1990s with the worldwide embrace of the World Wide Web.

The Web is the foundation of digital convergence technology. It combined computing (Web servers and PC-based Web browsers), digital communications technology (the Internet), and digital media (the Web included images as well as text and numbers) to create a new platform for human communication, collaboration, and information sharing. Complementary innovations such as email, instant messaging, and file sharing programs, such as Napster, contributed to the runaway success of Phase 1 convergence technologies.

The First Phase – Internet Computing

The early Web featured easy hyper-linked surfing and the use of images to enhance content and enrich and customize the user interface. The instant attraction of the graphics-rich Web, together with the obvious appeal of electronic mail, led to incredible growth in the Internet-based interconnectedness of our society and our economy during the first phase of digital convergence.

During phase 1, many Internet connections had very limited transmission capacity (also known as bandwidth). Most consumers had slow modem connections, and many businesses had low-speed links as well. The early Internet infrastructure, which includes the backbone networks that interconnect Internet segments and the access networks that bring the Internet to user desktops, was also unreliable, and it had a propensity for network congestion and delays. Use of audio or video was problematic at best. Most Web sites were limited to text, numbers, and highly compressed images – suitable for lowest-common-denominator telephone-modem–based Internet access.

As the Web evolved beyond being a tool for hypertext-based navigation of static pages, many Web-based applications emerged that were built on the bread and butter of corporate enterprise network computing:

database searching, transaction processing, and dynamically generated user interfaces. These applications were also built primarily with text, numbers, and compressed images, so narrowband Internet access worked fairly well. This included news and information services, online financial services, discussion forums, and e-commerce sites for buying, selling, or trading goods and services.

I've taken to calling this the Internet computing phase because much of the phase 1 technology was based on Web and Internet adaptations of computer and information technologies developed in prior eras. For example, many first-phase Web-based database applications were not drastically different, in terms of functionality and programming techniques, than the database-driven capabilities I developed on an IBM mainframe back in the early 1980s. The difference lies largely in the complementary use of images, some new hardware and programming technologies, and global access to the finished applications.

The Next Phase – Rich Media and Mobile Internet

Phase 2 promises to deliver some very different capabilities. First-phase constraints are rapidly being overcome, unleashing the incredible potential of the Web and the Internet as a global platform for media-rich forms of human communication, interaction, and information sharing. This next phase is characterized by two technological features – rich media and mobile Internet access – that promise to significantly enhance the impact of convergence technologies in our lives.

1. Rich Media

Broadband is a term that means "lots of bandwidth."* When technologists talk about broadband, they're usually referring to consumer and small-business Internet connections that are many times faster than the

* While there is no universally agreed upon definition of *broadband*, most would agree that a broadband connection must support a transmission rate of at least two hundred thousand bits per second.

slower, narrowband modem connections they replace. Cable modems give consumers and small businesses broadband connections; phone companies offer another broadband alternative, called Digital Subscriber Line (DSL). Broadband connections are finally reaching a critical mass among consumers and small businesses, and most larger organizations provide high-speed connections as well. The Internet infrastructure has also improved, eliminating many of the bottlenecks and much of the congestion that plagued the phase 1 Internet.

With Internet infrastructure improvements and broadband adoption gaining momentum, the use of rich media is becoming central to our Internet interactions. Internet telephony (or voiceover IP), streaming media, videoconferencing, Flash animation, and a host of other rich media technologies and services are becoming an integral part of many Internet interactions. We'll explore these and other rich media technologies as the book unfolds. We'll also examine the ways that information technologies are disrupting conventional entertainment industries (the largest single data type transmitted through the Syracuse University Internet connection during the past few years has been MP3 music files).

As I write this, I hear a young woman in the next room doing a rendition of "Let's Dance," a crossover country hit made popular by Trisha Yearwood. It turns out that my fifteen-year-old son's girlfriend, Amanda, is taking her turn singing at a Yahoo.com audio-enabled chat site, where people enter a chat room and take turns singing or playing instruments. This is pretty cool, and it offers a great example of the sort of creative applications that are possible on the new rich-media–enabled Web. My son rarely watches television.

2. Mobile Internet

> "It requires no great leap of the imagination to believe that the convergence of mobile communications and the Internet will produce something big . . . but it may take longer than you think."
>
> *Internet for a Mobile Generation: Executive Summary*
> International Telecommunications Union, September 2002

Wireless mobile networks are evolving in several realms including wireless LANs, Personal Area Networks (PANs), home networks, digital cellular networks, and broadband-enabled satellite constellations. We are seeing the beginnings of a robust, high-speed wireless network infrastructure that will eventually enable mobile Internet access at broadband speeds. You'll be able to get high-speed Internet access from a notebook computer at the airport; a Web-phone on a commuter train; or a Personal Digital Assistant (PDA) in your SUV.

While many believe mobile Internet is hugely important, I can't help but view it as a technology of convenience. In contrast, I believe rich media technology can and will empower people at the grass roots. Also, infrastructure and industry challenges may delay widespread deployment of high-speed wireless technologies, further constraining the near-term impact of mobile Internet. Still, this clearly is an emerging phase 2 development we need to explore.

Early in phase 2, many marketers and entrepreneurs drooled over wireless technology since they equated it with consumer-based commerce. They seemed to think that consumers would welcome a device that enabled them to check sports scores, stock prices, and the weather forecasts while being spammed with opportunities to buy stuff from the merchants and commercial chains in the immediate vicinity. They call this mobile commerce, or m-commerce.

I think they're wrong. The focus of mobile Internet, if it is to succeed, will be individual communication, collaboration, and information sharing.*

Mobile Internet technologies and industry challenges are explored in several chapters, and chapter 17 is dedicated to this important phase 2 technology.

* The ITU report mentioned at the beginning of this section also included a strong suggestion that legislation protecting consumers from potential m-commerce abuses may be needed. From the International Telecommunications Union, *Internet for a Mobile Generation: Executive Summary*, September 2002.

A Future Phase: Embedded Internet?

A future phase will see routine inclusion of embedded processors and Internet connections in a much wider range of devices and appliances.

An early example that was fodder for the high-tech hype machine was the Internet refrigerator. Think of a fridge with a built-in computer that could scan UPC codes from food products. It would know when you were running low on milk or eggs, and generate a shopping list or maybe even place an order via the 'net to a grocer near you. Whenever I pull this one out in class or in an executive workshop I get a chuckle.

Other, more realistic examples include sophisticated home and office appliances fitted with on-board computers and network connections, such as a home security consoles, lighting controls, and printers. These devices can be remotely controlled by end users and technicians (for configuration and diagnostic testing). Also, shared devices – such as vending machines, conference room video projectors, and arcade video games – will be equipped with embedded processors and 'net connections, enabling useful new functions.

Embedded Internet is still in its infancy and is largely a set of ideas, underlying technologies, and potential applications. From my standpoint it is an evolving, relatively low-impact vision that has me stuck in skeptic mode. Maybe I'll soon have a change of heart, and eventually take on the topic in a subsequent book. For now, rich media and mobile Internet are ready to make a difference to us all in the relative near term, so I'm focused on these realms of innovation. One or both may come to dominate the second phase of the digital convergence era.

Digital convergence follows other significant eras in the evolution of digital technologies, such as the mainframe, personal computer, and network computing eras. It promises to be as interesting and fruitful for the human race as any wave of innovation that accompanied any prior technological era.

Like many technological terms, "digital convergence" has alternative meanings and definitions in different circles.

- In the world of consumer electronics, digital convergence refers to the creation of a seamless end-user experience that results as entertainment technologies – television, stereo sound systems, video cameras, etc. – all make the move to digital.
- In the telecommunications community, digital convergence refers to the transmission of voice, video, and data across a single digital telecommunications infrastructure.

I believe my definition encapsulates these other notions while providing a fairly simple framework and more effectively illuminating the power and promise of this overarching technological phenomenon.

Individual Empowerment and Corporate Challenges

Some believe that when it comes to your average business professional, any detailed knowledge of technology is either essentially unobtainable or a waste of time. For example, Christopher Locke wrote the following in *Gonzo Marketing*[*] for his audience of business executives and marketing professionals: "Bandwidth is an enormously complex concept, one which you are probably not congenitally predisposed to understand." While I assume that Locke has his tongue firmly planted in his cheek (and I agree with most of what Locke writes), I wholeheartedly disagree with the all too common notion that business pros can't or don't need to know about technology.

I believe very strongly that it is important and helpful for individuals to gain some basic understanding of how technology works and how it is evolving. I am totally committed to this. And I know full well that the knowledge and insight I seek to share is important for individuals who

[*] Christopher Locke, *Gonzo Marketing*, Perseus Press, 2001.

are looking to recognize and seize opportunities that new technologies enable.

I have explained bandwidth, broadband, and a host of high-tech concepts and technologies to a wide variety of business students and business professionals, and almost all are fully able to understand and appreciate the basic concepts. What's more, most of the students and executives I interact with quickly grasp how and why this is empowering. In the age of digital convergence there is no doubt that knowledge is power.

From a technology standpoint, this book builds on the information and insight presented in the first book in a few ways. For one, it acts as an update, providing information and insights about emerging application technologies that enable new forms of human interaction. It also offers a more in-depth analysis of the factors that affect the evolution and convergence of underlying technologies, and it focuses considerable attention on the promise, potential, and pitfalls of convergence technology for creators, collaborators, and consumers.

The book will explore the continuing evolution of the convergence phenomenon and explore the empowering impact on individuals – while stimulating you to think through how you might leverage new technologies to benefit your own life and career. But there is something else that we'll explore: the impact on large corporations, who have been the biggest beneficiaries in prior technological eras, but who are challenged by this latest technological phenomenon.

Many corporations are deploying digital convergence technologies with some success. These are typically internal systems that enhance corporate communication and productivity, and external initiatives to reach out to customers, suppliers, and business partners. Unfortunately, many large corporations – which too often cling to command-and-control management and are blinded by the aggregate view of "the firm" – often fail to recognize the potential of convergence technologies to empower employees working in the trenches. Instead, they fight the grassroots momentum that is inevitable as the convergence phenomenon takes hold. This is discussed in more detail in chapter 4.

Many corporations also bring a mass media, mass market mentality to new convergence-based platforms. They don't fully recognize that convergence technologies support a new electronic environment that facilitates niche markets, narrowcasting, special interests, and individual, group, and community-of-interest interactions. This is a recurring theme that will get some special attention in chapter 3.

And finally, the increasingly significant societal, economic, and technological shifts that digital convergence facilitates have created a new tension: digitally-empowered creators and collaborators versus the near-term, bottom-line strategies and tactics of corporate technology, communication, and media providers. We also will explore this telling tension as the book unfolds.

My Perspective

No author or consultant out there can give you a one-hundred-percent accurate depiction of what is happening as new technologies evolve, and no one can predict the future. It's all just changing too fast, with too many variables and unknowns. What I can and will do is explore the technologies and trends that are reshaping how we interact, communicate, collaborate, and share information. I also hope to stimulate you to reflect on the digital convergence phenomenon and identify potential opportunities from your own personal and professional standpoint.

This treatment of digital convergence and its impact is not backed by systematic, quantitative research. It is one man's view – presented to inform and stimulate you and your views. I think I offer a useful perspective on evolution of technology and its impact.

Much of my "research" has been done in the context of:

- my work in the trenches, where I strive to make technology work for myself and for my university, department, workgroup, clients, students, friends, and family

- the work I've done for various trade publications, which has given me the opportunity to deploy and analyze many new digital convergence technologies
- the fairly extensive network of colleagues I've interacted with and interviewed over the years – colleagues that include practitioners, researchers, and business executives in and out of high-tech companies

My experiences at Syracuse University have also enabled me to view firsthand the significant, albeit somewhat strained, impact that new technologies have on large organizations. Syracuse University has some 3,000 to 4,000 full-time employees – not a mega-corporate entity like an Exxon or General Electric, but still a sizeable collection of individuals working together to achieve a set of individual, departmental, and institutional goals.

Today's university is out in front in terms of Internet and broadband penetration,* and I'm convinced that this environment is a model for effective use of new technology. The university makes available significant funding for new technology, and it has a healthy share of intelligent, creative individuals intent on acquiring and leveraging new technologies. I will share with you many examples and insights from my university experiences.

I've also witnessed the enormous impact that new technologies can have and have had on individuals. I've been a technology professional for many years, and as a supervisor, I've always been a hands-on, "player-coach" technology manager. This experience has given me an in-depth understanding of not only the promise of new technology, but also the significant dark side of technological unreliability, instability, and user un-friendliness.

* For example, a study from the Pew Internet & American Life Project found that 86 percent of college students have gone online, compared with 59 percent of the general population; and 74 percent of college Internet users go online for fun, compared to 64 percent of all Internet users. From *The Internet Goes to College*, principal author Steve Jones, September 15, 2002.

I teach undergraduates, many of whom leverage technology in interesting and instructive ways. I am also a baby-boomer who teaches other baby-boomer executive- and continuing-education students. We grew up in the television generation, well before digital technologies took hold. These experiences have further broadened my understanding of the role of technology in the lives of creative, competent people.

Finally, I have a fifteen-year-old son, which may not seem like a solid qualification to some. But his generation, unlike mine, has grown up on the technologies of digital convergence, and my high-tech son has exposed me to some of the innovative ways that this generation uses technology. It has been fascinating.

I've worked hard to make the technical content of this book accessible. This is crucial so far as I'm concerned if this book is to have the impact I believe it can have. That doesn't mean I have dumbed everything down to the point that I avoid all buzzwords and instead focus on weak analogies and comparisons (e.g., the Internet is like a superhighway with several lanes of traffic, blah, blah, blah). I believe I can present the technologies and technical issues at a level that offers a reasonable understanding, with a level of detail that many find just right. I've gotten plenty of feedback from the first book that bolsters my confidence in this regard.

This book is meant to be a field guide and updated primer on the digital revolution as we move into a new phase of technological infrastructure and capability. I'll try my best to cut through the hype and explain how the technology is really playing out in our society and in our lives. After all, the reality is often far from that depicted on television commercials offering a view of "riding the light."

Here's how I've laid out the rest of the book:

Part II will explore the enriching and empowering impact of digital convergence technologies on individuals and on our corporations and institutions. Here you'll find my thoughts on the intersection of convergence technology and human endeavor. In this section I hope to stimulate your thoughts about the impact of technology on our lives.

Part III will provide an introduction to the forces that drive and constrain the evolution of the digital convergence technologies. I'll try to explain the layering of convergence technologies and review some of the technological and non-technological factors that influence the evolution and convergence of underlying computing, communications, and digital media technologies.

In Part IV we will look at some emerging application technologies that people like you and me will use to interact, communicate, collaborate, and share information in new ways.

Part II

The Impact of Digital Convergence

This section presents a series of ideas and insights, taken from an individual, human perspective, that highlight what I believe will prove to be the most significant impacts and outcomes as the convergence phenomenon unfolds. In the final chapter of this section, I'll tie together the ideas presented in the next three chapters with a stab at a list of the entities that I believe will be the "big winners" in the digital revolution. I hope this stimulates you to think through what convergence means to you given your own professional and personal perch, while providing encouragement and motivation for you to effectively pursue new technologies.

The automobile changed individual lifestyles and work habits tremendously. For example, it enabled millions to move outside the concrete jungle, creating a donut of suburbs surrounding decaying inner cities. There are similar opportunities for lifestyle and workstyle enhancements in the age of convergence. Only this time around we'll be able to bring the world and its opportunities to us, while bringing ourselves to the world – without ever touching the gas pedal.

2. CREATOR-COLLABORATORS

Much has been written in the business press about the organizational impact of new technology. Beyond the rather generic and variously defined e-commerce and e-business, typical topics are enterprise-wide opportunities to use technology in enterprise resource planning; supply-chain management; and customer relationship management. Some great books by professors, executives, and consultants have also explored the impact of technology on large organizations. These focus on technology-facilitated business and organizational concepts such as human capital, the networked organization, and business arrays. However, they typically pay little attention to the perspective of the individual when it comes to the overall impact of the digital convergence phenomenon.

Many of the sources that do explore the impact of new technologies on individuals focus strictly on a consumer-as-target agenda with topics such as e-marketing, new media, customer service, e-tailing, and other resources and services aimed at consumers. Then there are the how-to books, articles, and online tutorials that explain how to perform a specific set of tasks to address the needs of individuals working "hands on" with a particular technology application. These texts offer little to help individuals understand the big-picture phenomenon that sees convergence technologies empowering individuals in a manner never before seen in any prior technological era.

I've found only a few texts that seem to really "get it" when it comes to capturing the impact of new technology from an individual, human perspective. They include:

- The Cluetrain Manifesto (at www.cluetrain.com) and the subsequent book by the same name.* This work provides a stimulating

---

* Rick Levine, Christopher Locke, Doc Searls, David Weinberger, *The Cluetrain Manifesto: The End of Business as Usual*, Perseus Books, 1999.

summary of the enabling impact of the Internet and the Web on today's consumers, including "Joe Sixpack."

- *Brand You 50,*[*] Tom Peters' work on personal branding from his "Reinventing Work" series. This book offers a spirited call to action for today's technology-enabled knowledge worker who can become what Peters calls the "Survivor-Thriver" in the "New World of Work."
- The online writings of Simon Buckingham at www.unorg.com. Simon details a phenomenon he calls "unorganization," and profiles the rise of "technological capitalism," where individuals become increasingly empowered by technology – to the point that they can and do profit directly.

These sources focus considerable attention on the individual, who I believe is at the heart of the digital convergence phenomenon. My efforts are an attempt to complement these and other writings by:

- exploring the manner in which new technologies empower the individual
- highlighting the ability of individuals to profit from technology use by introducing a new entity called the creator-collaborator (defined below)
- linking the rise of the creator-collaborator to the digital convergence technology phenomenon
- offering knowledge and insights to help creator-collaborators who want to aggressively use new technologies
- clarifying the power of communities of creator-collaborators to challenge corporate control of our technology and our lives

[*] Tom Peters, *Brand You 50*, Knopf, 1999.

Let me begin by introducing you to creator-collaborator Phil Leigh. Phil is a very nice fellow, and he's very sharp – and I'd like to make an example of him.

When I first met Phil he was a stock analyst with Raymond James & Associates. This is a full-service investment firm that is part of the Raymond James Financial family of investment firms, headquartered in Tampa Bay, Florida (where the Raymond James brand name adorns the football stadium of the Tampa Bay Buccaneers).

As an analyst, Phil followed the activities of companies in the digital music, digital video, streaming media, and related marketplaces. He also followed the electronic conferencing market, which includes companies offering Internet videoconferencing and Web-conferencing services and products. He studied small startups at the fringe of the online digital entertainment and Web conferencing movements on up to giant technology providers like Sony and Microsoft.

Phil, who is keenly aware of the power of new convergence technologies, launched an electronic presence to facilitate his work. As a result, Phil became a very visible expert in his field. I interviewed him in the summer of 2002, and he had a very interesting story to tell.

Phil was first contacted by an Internet streaming audio site venture called Radio Wall Street that was looking for knowledgeable, talented individuals to provide compelling content. He soon had a regular Radio Wall Street gig as analyst and interviewer. Phil used a listserv to send out email announcements of upcoming guests and to periodically disseminate market research reports on topics such as online music, streaming media, and Web conferencing. He eventually moved to a new site, ON24.com, when Radio Wall Street folded.

Over time Phil developed quite a following. He told me he knew he was on to something when digital media companies started pursuing him to get on his interview calendar. Of course, you've probably never heard of Phil. Although he is occasionally interviewed on television and is a staple of technology-trade-rag beat writers, he is far from a household name.

When I spoke to him, he was unable to point to any direct, quantifiable link between his online presence and his bottom line. But Phil knew full well that this experience had made him a better stock analyst. He had much better access to digital media industry insiders, and he received valuable feedback from informed audience members that helped him to refine his thinking. Phil also indicated that this venture had had a huge impact on his life and career – and he knew it was something he was compelled to continue. At the time he was extending his electronic presence with integrated PowerPoint slide shows and downloadable MP3 recordings.

In the summer of 2003 Phil left his position with Raymond James to strike out on his own as a freelance provider of digital media technology and marketplace information and insights. His new electronic presence is called *Phil Leigh's Inside Digital Media.**

Phil Leigh is one of a new breed of individuals who aggressively leverage new technology to enhance career and/or lifestyle. There are countless stories of digitally empowered individuals who successfully leverage new technologies to move their lives and their careers in exciting new directions. While Phil's leap was sort of a no-brainer, given the fact that he was utilizing those technologies that had been the focus of his research and analysis, countless similar stories play out in a range of businesses contexts. Check out the heating, ventilating, and air-conditioning super-site, HVACTV.com, if you don't believe me.

There are many variations on this theme of individual empowerment, and we'll explore some other examples below. These are the stories of what I call the creator-collaborator.

Introducing the Creator-Collaborator

You've undoubtedly heard some of the buzz in the press surrounding *B2C* (business-to-consumer) and *B2B* (business-to-business) e-commerce

* Phil's new Web site is at http://insidedigitalmedia.com.

activity. Based on what is written and aired in the broadcast, business, and trade media about the business impact of new technology, you would likely conclude that the impact of digital convergence is exclusively focused on these two entities. But the digital convergence phenomenon has given rise to an important new entity: the creator-collaborator.

Business (B):	A commercial enterprise or establishment.*
Consumer (C):	One who acquires goods or services; buyer.
Creator-collaborator (C'):	One who aggressively uses technology to enhance lifestyle and career.

The creator-collaborator is becoming a powerful entity in our society and our economy. Typically a knowledge worker or tech-savvy consumer, a creator-collaborator uses technology to create new resources or interactions and to collaborate with others. All creator-collaborators seek to profit from individual technology-based initiative and effort, either professionally or personally – or both.

With this new convergence-enabled entity now defined, we can more accurately depict the business impact of digital convergence. Figure 1 shows some examples.

Creator-collaborators are very different from conventional consumers who sit back, waiting for the mass media message that tells them what technologies to like and use. Creator-collaborators proactively assess and explore their technological options, and create new ways of approaching life and work in the process.

---

* Definitions for "business" and "consumer" are from the *American Heritage Dictionary*.

Figure 1

Transaction	Business Scenario	Technology or Resource
B2C	E-tailers and click-and-mortar businesses	Amazon.com
B2B	Business portals and exchanges	Foodservice.com
B2C'	Early adopters of empowering technology	Broadband Internet services
C'2B	Free agents and consultants on-line	Freeagent.com
C'2C'	Professional online communities and virtual teams	Slashdot.org
C'2C	Niche-oriented products and services	eBay

While the economic impact of creator-collaborators is significant and growing, the creator-collaborator is much more than a business entity. Creator-collaborators are an important force for social change (both positive and negative), and communities of creator-collaborators are creating technologies, media content, knowledge, and interactions that go well beyond what many well-funded corporate providers can muster.

I'll bet that you or someone you know is a creator-collaborator, leveraging new digital technologies to advance personal or career objectives. It's becoming rather routine in most parts of the developed world. Odd Todd and Phil Leigh offer compelling examples of what creator-collaborators can accomplish. Let me share a few more modest cases taken from my own personal experience:

- The friend of a friend who owns a local towing business with his father. He got on the Web a few years ago – he was one of the first on his block to get Roadrunner, the AOL Time Warner broadband option offered in our area – and he quickly surfed his way to a business relationship with an Italian car parts dealer.

- The friend of a teacher my wife knows who started wheeling and dealing on eBay. Her efforts paid so handsomely she decided to scrap a solid day job and go full-time into online auction sales.
- The Syracuse, New York, gardener who was so excited with her gardens (yes, we do have beautiful summers here), that she decided to take digital camera and a little time to create a Web site to share her floral successes with the world. She gets global visitors who offer advice and encouragement. She also visits sites like gardenweb.com to share information and camaraderie with fellow gardeners from across the globe.
- My son. He's into Starsiege Tribes, a team-based, first-person shooter game (he gets to shoot and kill other players). He and his comrades form "tribes" that are a registered with the Official Gaming League (OGL tribes have frequent matches with other OGL tribes, and the OGL ranks both tribes and players). Audio feeds and text-based chats enable communication during matches. My son has recruited players from England and Singapore, and his tribes have reached the top ten on more than one occasion. He encounters politics, loyalties, and human conflict. He's developing skills that I believe will serve him well once he enters the real-life world of work in a few more years.
- Ebenezer, my pen pal from Ghana. He contacted me after reading some of my articles in *Network Computing* magazine online. He recognizes the power of new technology and is seeking to become a technology professional. He has asked for me to send him manuals, trade magazines, or whatever I can get my hands on that would help him learn more about technology. He has sent me numerous emails, pictures, and online greeting cards. He has detailed the efforts of the government of Ghana to move in the

> direction of a high-tech economy. Now he's asked for some financial assistance to take some computer certification courses. I think he has several pen pals, and I must admit I'm impressed.

Pradeep Jotwani, president of the Hewlett-Packard Consumer Business Organization when I interviewed him in 2001, used the term "lifestyle enrichment" to refer to the personal empowerment that technology enables. He also attributed many innovative individual successes to the work of what he terms "early adopters."*

Pradeep is a proven leader in one of the world's largest tech companies, and he and his staff at HP have been a great resource for me during the writing of this book. But his characterization captures only a portion of the reality when it comes to convergence-enabled innovation. And I believe it comes up short when gauging the overall impact of new technologies for so many creative, energetic individuals who can do so much more than send faster mail, get better TV, and buy, buy, buy the stuff they are bombarded with on "Must See TV."

It is true that many convergence technologies, such as digital cameras and digital television set-top boxes, are squarely aimed at passive consumers who may want to entertain themselves, share family memories, or make life more enjoyable. Creator-collaborators use convergence technologies not only to enjoy free time, but to learn about life and work, enhance job performance, and generate income.

There is a continuum when it comes to creator-collaborator behavior. At one end of the spectrum, you have what appear to be classic consumers with an added technology-enabled agenda; at the other end you find entrepreneurial creator-collaborators who use technology to bootstrap promising new ventures. I'll profile five types of creator-collaborators to illustrate some of variation when it comes to how individuals pursue technology to enhance lifestyles, careers, and business ventures. I'll then itemize a set of characteristics to further flesh out this important new entity.

---

* My July 2001 interview with Pradeep Jotwani is available in streaming audio format (http://whitman.syr.edu/connections).

Types of Creator-Collaborators

1. Lifestyle/workstyle enrichers

This entry-level category includes those who use technology to share experiences with friends and families, create new entertainment opportunities, pursue hobbies, or electronically engage colleagues, business partners, or customers. Lifestyle/workstyle enrichers are actively engaged and growing with the technology. The outcomes are usually modest, but many learn a great deal and experience a significant return on investment – and then move on to more ambitious creator-collaborator activity.

2. Creatives

Creatives launch more serious efforts to identify and implement technology-based tools, resources, and services. Most creator-collaborators in this category work to enhance a business or career, although some pursue creative activity outside of work. This group includes salespeople who aggressively use the Web to identify prospects or use mobile technologies to enhance communication and on-site sales presentations; hobbyists who develop some serious hobby-related digital resource or interaction; worker bees who organize work-related online forums or create professional content to gain recognition and gain increased access to expertise; or workgroup managers who actively seek out and deploy services, technologies, and resources that can deliver increased sales, decreased costs, or enhanced productivity and efficiency.

3. Free agents

Millions of knowledge workers are joining a self-employment movement called the free agent movement. While a primary stimulus for the emergence of the free agent movement has been the erosion of corporate stability and loyalty, digital convergence has also been a crucial enabling factor. Daniel Pink, author of *Free Agent Nation*,* articulated this in an interview published on the tompeters.com Web site:**

> "Certainly technology has been another reason [for the beginning of free agency]. Free agency is essentially Karl Marx's revenge, because workers can now own the means of production. In the industrial economy, the tools you needed to create wealth were large, expensive, and difficult for one person to operate. But now the tools – such as a laptop – are small, affordable, and easy for one person to operate."

As the technologies of convergence continue to provide new functionality at reduced cost, expect this group of creator-collaborators to remain in the forefront.

4. Entrepreneurs and intrapreneurs

These are the folks, like Odd Todd (oddtodd.com), who take it to the next level and generate significant additional income. This is often accomplished outside of any corporate setting; for example, a fledgling business venture or a hobby that becomes a significant business opportunity. Or it may be a grassroots, job-related technology initiative that leads to significant sales leads, customer satisfaction, or cost savings. This category includes innovators, such as Phil Leigh, who may not see a direct link to the bottom line when starting up – but who nonetheless take risks to implement a rewarding new venture.

5. Grassroots geeks

This category comprises the legions of technically adept individuals who work collaboratively – often outside of any corporate structure –

* Daniel Pink, *Free Agent Nation*, Warner Books, May 2001.

** You can find the complete interview on the Web site (http://www.tompeters.com/cool_friends/content.asp?id=26).

to create new technologies and improve existing technologies that enhance lifestyles and careers. This includes the armies of programmers and hackers who create free software such as Napster and Linux, which anyone is free to download and use. It also includes individuals in grass-roots, online communities such as the TiVo community forum – which provides, among other things, online help for people who wish to add a disk drive to their TiVo personal video recorder (PVR) to increase the available hours of recording capacity.

Creator-Collaborator Characteristics

Now that you have some sense of what a creator-collaborator is, let's look at a few telltale characteristics of this rising force in our economy and society.

1. Creator-collaborators often blur the line between work and play.

Creators-collaborators recognize that uses of technology for work and for personal gain are overlapping. What you learn about technology in the workplace – e.g., accessing and interacting with Web resources and services, using email and discussion forums, creating Web sites, or using wireless technology – is often relevant in pursuing personal objectives. Likewise, expertise or technical skills developed in individual pursuits – e.g., interacting with remote contacts who share a special interest such as a hobby, hooking up a router/switch to share a broadband cable or DSL connection, creating a family Web site, or buying and selling products on an auction site – are often transferable to the workplace.

2. Creator-collaborators are often focused on special interests.

Creators-collaborators typically leverage what I believe is one of the key strengths of digital convergence technologies: an ability to facilitate interaction and information sharing around a special interest. The Internet and the technologies of convergence are less about mass markets and mass media, and more about niche markets and individual, group, and community-of-interest interactions. If you're interested in the Super Bowl, you'll use conventional broadcast technology to access the live event. If you're interested in something far less popular, but still meaningful for you, you'll turn to the technologies of convergence.

3. Creator-collaborators are used to overcoming the rough edges of new technology.

Creator-collaborators are typically unafraid of technology and are willing to live with or work around its rough edges. Passive consumers typically lack the incentive or patience to deal with those sometimes frustrating aspects of aggressive technology use, but creator-collaborators are willing to put in some time and effort and work through the frustration to gain the benefits that new digital convergence technologies promise. This reminds me of that age-old slogan from athletics: "no pain, no gain."

4. Creator-collaborators sometimes break the rules.

Creator-collaborators are often willing to bend and sometimes break rules that may seem to be outdated as new technologies enable new forms of human interaction and communication. They may share copyrighted entertainment content. They may surf the Web while at work, checking out sites they may use to enhance their lives in ways that may or may not be work related. In cases where employees seek to circumvent corporate control over technology, content, and interactions, it is usually creator-collaborators who create, distribute, and use the technology.

What About Consumers?

Consumer product companies will provide improved service and support for their products, using the Web and other technologies that enable additional consumer touch points. Consumers will also continue to see many more entertainment and education options. The Web, digital television, mobile communications and computing devices, and a host of other technologies – from digital cameras to the Xbox – are providing a growing range of exciting alternatives for individual consumers.

Consumers are also involved in a new Internet-enabled conversation that yields insight about all kinds of consumer-oriented products and services (the Cluetrain Manifesto, mentioned earlier and presented in more detail in the next chapter, focuses on this notion). Consumers will surely have more power, they'll be better informed, and they will clearly benefit from convergence technologies. But creator-collaborators will thrive, prosper, and flourish.

The next chapter will place the rise of the creator-collaborator in the context where the creator-collaborator has the greatest impact: the realm of information, concepts, ideas, and knowledge.

3. MIND OVER MATTER

My aim in this chapter is to give you a piece of my mind.

I don't mean this in the classic sense of cruelly and aggressively laying out a litany of your faults or blithering at you with anger and disdain, but rather, gathering together my thoughts on a particular topic, and sharing them with you through the printed word. You are scanning the letters and words, and together they provide some specific meaning. They generate some mental images, and perhaps some feelings and reactions – and ultimately a new understanding. This is really a rather amazing transaction, but it's such a routine part of our human experience that we don't even really think about how wonderful and powerful it really is.

Experiences of the mind – such as the experience of reading these pages – shape your being. This is what sets us apart from other life forms. And it is something we often take for granted.

I was attuned to the power of my mind over my being during my college years when I took a course in Buddhism and experimented with meditation. Unlike western religions, which are highly conceptual and devotional, Buddhism relies heavily on practice. Zen Buddhism, in particular, caught my attention.

The goal of Zen practice is to achieve enlightenment, which is emptiness – an emptiness devoid of the thirsting that human concepts and ideas engender. In the words of Zen Master Dogen Zenji, "you must suspend your attempts to understand by means of scrutinizing words, reverse the activity of the mind which seeks externally, and illuminate your own true nature."*

---

* Translated essays by Francis Dojun Cook, *How to Raise an Ox,* Center Publications, 1978.

Master Dogen Zenji also describes meditation, known as zazen:

> ". . . when your body posture is correct, breathe in and out [once, deeply]. Sway left and right [several times] and then sit firmly and resolutely. Think the unthinkable. How do you think about the unthinkable? Do not think. These are the essentials of zazen."

I experienced zazen, and it was an eye opener. I constantly have thoughts churning through my mind, and the experience of focused relaxation that stops that flow of concepts, ideas and images – just throwing them away as soon as they come to mind – was both fascinating and rejuvenating.

The world we experience is a physical world, but our experiences as intelligent humans – and the experiences we desire – are increasingly shaped and filtered by the concepts, ideas, and mental images we carry around in our hyperactive human brains. Yes, we eat and sleep, exercise, make love, and engage in other physical, worldly activities. But it is what happens in our minds that seems most important in the day-to-day living of life.

We live in a hyperconceptual human society. We generate ideas and communicate concepts. We understand facts. We listen to reason. We check the weather by doing more than peeking outdoors. We have sophisticated and complex conceptual structures for committing to lifelong adult relationships; integrating our children into society; engaging in work activity to gain shelter and food; celebrating life with loved ones on national and religious holidays; and forging a lifestyle. Concepts, ideas, and mental images drive our lives and our desires.

Most of us blindly accept our conceptual realities, and seem to generally enjoy our lives, though most of us, at least at one time or another, question this existence.

What does this have to do with digital convergence? There are a couple of huge and unavoidable implications of "mind over matter" when it comes to new technology and what it means to consumers and creator-collaborators.

- First, consumer consciousness has become a playground for marketing and advertising moguls who have honed the use of conventional mass media. The resulting consumer-oriented society is finally seeing a challenge from grassroots use of convergence technologies.
- Second, most of us recognize that our mind is the workhorse that pays the bills and offers the best opportunity to get ahead in today's knowledge-based economy. Concepts, not muscles or shoe leather, most often shape our work and drive profits and promotions. Significant individual opportunities will continue to present themselves as new convergence technologies empower us to leverage our unique knowledge, expertise, and experience.

Let's explore these two notions a little further as they relate to the reality and promise of digital convergence.

Mass Media and Mass Markets

> "Beyond the sphere of local primary contacts of the citizen, the mass media provide most of the 'reality' for the perceived world."
>
> Leslie Moeller, *Beyond Media**

Mass media technologies have been used for several decades to disseminate concepts, ideas, information, and mental images to the masses. During the twentieth century, the big four – newspaper, magazine, radio,

* Leslie Moeller (editor), *Beyond Media* (second edition), Transaction Books, 1988.

and television – provided highly organized means of not only informing, but also influencing public perceptions.

The corporate mass marketing engine is keenly aware that we soak up new ideas, concepts, and mental images. And it recognizes that these things have a huge impact on our behavior. The mass media have been the vehicle of choice for building demand for product consumption. Over the years, the effort, analysis, artistry, and funding with which corporations pursue this activity have been raised to what seem to be unbelievable levels.

Cost of a 30-second ad spot:

$2.3 million

Super Bowl XXXVIII*

When you buy a soda, what do you buy? Do you buy a Coke or Pepsi, or maybe Mountain Dew? Or maybe a Dr. Pepper? These are brand names. They're developed and promoted to make you feel a certain way about these beverages, beyond your reaction to the taste. What's the Pepsi tag line? The Joy of Pep-si! I'm sure you've heard Britney sing it. Pepsi is a very strong brand.

* Suzanne Vranica, "Bud Bowl Ads Score Points," *The Wall Street Journal Online*, February 2, 2004 (http://interactive.wsj.com/documents/retro_superbowl04.html?mod=home_inside_today_us); Stuart Elliot, "During Breaks in Game, Satire and Silliness," *New York Times*, February 2, 2004 (http://www.nytimes.com/2004/02/02/business/media/02adco.html?adxnnl=1&pagewanted=all&adxnnlx=1075743584-58nyBxU+qOzQvyVGpitH+Q).

> "The astronomical growth in the wealth and cultural influence of multinational corporations over the last fifteen years can arguably be traced back to a single, seemingly innocuous idea developed by management theorists in the mid-1980s: that successful corporations must primarily produce brands, as opposed to products."
>
> Naomi Klein, *No Logo*[*]

Branding has become a corporate religion that seeks to embed perceptions in the minds of consumers that will ultimately lead to a vastly improved bottom line. The target of branding is the "consumer." No one likes to be labeled, and the term consumer has become for me a gross, simplistic conceptualization – that of a singular product-hungry group of individuals that is a gold mine for successful mass market merchandisers.

Through my fifteen-year-old son, I have seen the results of aggressive brand building. When I was his age we wore tee-shirts that indicated our lifestyle, interests, or aspirations with college and sports logos, funny sayings, and rock band concert schedules. Now, at least in my son's circles, a swoosh or Aero or some other acceptable logo is an absolute must.

I'm also a sports fan, and I've seen sports become inundated with branding efforts – whether it's the PGA Buick Open, the Tostitos Fiesta Bowl, or the NASCAR Winston Cup. I'm sure you can quickly think of many other examples of increased brand influence.

During the last several decades, corporations have muscled in on our mindshare to create a culture that generated very profitable commercial activity. Huge corporate empires were created in the process. I believe the pendulum is now beginning to swing back in favor of the individual – and the empowering technologies of digital convergence comprise a major push in that direction.

[*] Naomi Klein, *No Logo*, Picador, 1999.

The Cluetrain Manifesto (www.cluetrain.com) and the follow-up book, *Cluetrain Manifesto: The End of Business as Usual,*[*] attempt to clarify the impact of the Internet on markets. The book offers a pointed analysis of the Internet as a consumer empowerment zone, not a mass market medium. According to the *Cluetrain* authors, today's emerging marketplace is an Internet-enabled conversation.

Here are a few of what I think are some of the more interesting Cluetrain Manifesto theses to give you a taste for what this is all about:

#1 Markets are conversations.

#6 The Internet is enabling conversations among human beings that were simply not possible in the era of mass media.

#12 As a result, markets are getting smarter, more informed, more organized. Participation in a networked market changes people fundamentally.

#18 Companies that don't realize their markets are now networked person-to-person, getting smarter as a result and deeply joined in conversation are missing their best opportunity.

#39 The community of discourse is the market.

#40 Companies that do not belong to a community of discourse will die.

This is pretty edgy stuff, and students in my undergraduate and executive courses have mixed reactions. However, at some level almost all of my students believe that the authors are onto something.

I agree with much of what the Cluetrain Manifesto authors say – but I believe the ability of new technologies to empower individuals goes well beyond conversation in the marketplace and in the corporation. Yes,

[*] Rick Levine, Christopher Locke, Doc Searls, David Weinberger, *The Cluetrain Manifesto: The End of Business as Usual*, Perseus Books, 1999.

consumers are empowered, and the power of the mass media to shape our lives and our desires is now challenged by convergence-enabled market conversation. But beyond that, consider the creator-collaborator who has new tools to enable her to effectively forge a new workstyle and lifestyle, and build her own personal brand in the process. This is very possible, especially if she works in the realm of individual impalpables.

Impalpables

On December 3, 1997, Alan Greenspan, Chairman of the Federal Reserve, delivered a speech to a group of Syracuse University students.* Unlike most other Greenspan speeches I've seen either on television or in person, this one focused very squarely on the impact of technology and education on the global economy and on the operation of today's businesses. Among Greenspan's comments:

> "[The] striving to expand the options for satisfying the particular needs of individuals inevitably results in a shift towards value created through the exploitation of ideas and concepts – or more generally information – from the more straightforward utilization of physical resources and manual labor."
>
> "Thus it should come as no surprise that over the past century, by far, the largest part of the growth in America's real gross domestic product is a result of new insights – and more broadly new information – about how

* This speech is available in Windows Media format in the "Archives" section of The Whitman's School's *Business Connections* Web site (http://whitman.syr.edu/connections).

> to rearrange physical reality to achieve ever higher standards of living."

> "The amount of physical input into our real GDP, measured in bulk or weight, has contributed only modestly to economic growth since the turn of the century."

As you might surmise from the above excerpts, Greenspan focused considerable attention on the rapid growth in "almost wholly conceptual good and services," which he dubbed "impalpables." He also talked about the increasingly important role of knowledge workers and new technology. These two concepts are intertwined and are at the core of what I believe are the most significant fundamental changes, from a business standpoint, that new digital convergence technologies enable.

If you look now at the most successful new-economy products and services, it is the impalpable industries – such as financial services, entertainment, and education – which are undergoing the most dramatic transformations. This is because most impalpables can be created, stored, managed, and transmitted digitally. The digital convergence technological phenomenon is an incredible fit for impalpable industries.

When Greenspan talked about impalpables, he included the travel industry and medical services. This got me thinking. With travel, there are some real physical components like jets and fuel and runways. A ticket is simply a conceptual place-holder for the physical reality that will eventually occur. Likewise, medical services are impalpable in some sense, because there is no physical hard good that the customer holds onto as an outcome. But the definition of impalpable in my American Heritage dictionary is "Not perceptible to the touch; intangible." While a teeth cleaning is an intangible service according to Greenspan, surely when the dental hygienist works that tartar off and picks and prods, making my gums bleed, it is definitely perceptible to the touch!

This led me to think about impalpables that somehow contribute to the creation of physical goods and services, including so-called hard goods – things like strategic plans, sales projections, and product design specifications. The reality is that almost any physical product or service is surrounded by a number of supporting conceptual goods and services. I call these individual impalpables, since they are most often developed and shared by individuals, workgroups, and teams of individual human beings (Figure 1).

Figure 1. Individual Impalpables: A Sampling

marketing plans	mission statements
technical specifications	sales projections
design drawings	contracts
Web sites	research reports
white papers	quality surveys
grant proposals	software
human resource policies	business plans
investment advice	bid specifications
financial reports	training modules
advertising concepts	product descriptions

In the world of hard goods and services the greatest impact of convergence technologies is not in the actual physical production of the end-product. The larger impact is in the technology-enabled creation, use, management, and sharing of a multitude of supporting impalpables.

This notion was solidified in my mind when I read an article by Jonathan Rauch titled "The New Economy: Oil, Computers, and the Reinvention of Earth."* This is an account of technology use in the oil business, which many perceived to be an old-economy industry. He wrote the article after an extended visit to an active oil derrick in the Gulf of Mexico. Rauch profiled the extensive use of information technologies to ex-

* Jonathan Rauch, "The New Economy: Oil, Computers, and the Reinvention of Earth," *Atlantic Monthly*, January 2001.

tract oil from the earth – at one point indicating that the drilling operation resembled a "computer with a drill attached." The article effectively tied the oil industry to the information and technology-based new economy. Among Rauch's insights:

> "Most people understand intuitively that the essential resource in Silicon Valley is not magnetic particles on floppy disks, or hard drives, or servers, or lines of code or bits of data; it is human ingenuity. The oil business cannot be information-based to the extent that the software industry is, but increasingly it is dominated by the same principle. Knowledge, not petroleum, is becoming the critical resource in the oil business; and although the supply of oil is fixed, the supply of knowledge is boundless."

This issue of *Atlantic Monthly* also happened to have an article on "Why McDonald's Fries Taste so Good," and believe me, the next time I drove through a McDonald's – to get a Number Two with a Coke for me, and two Filet-O-Fish sandwiches (plain) and a McFlurry with M&Ms for my son – I saw a different restaurant.

The new McDonald's I saw was not just a hamburger joint with a gaggle of teenaged burger flippers and a greasy manager. I saw a computer-assisted automated fast-food delivery system. Check it out next time you drive through or stop by. You can see the order confirmation monitor, check out the LCDs behind the beverage counter, and see the conveyor-belt–assisted automated drink supplier. Clearly, the design of the McDonald's food delivery system is built by an army of knowledge workers and a massive foundation of supporting impalpables. Ditto almost every other industry in the developed world today.

For decades, knowledge workers have been the driving force of our economy, providing the ideas and insights that generate value from raw resources. In the age of digital convergence, knowledge workers (especially those who are creator-collaborators) will increasingly use new technologies to enhance individual and organizational performance.

The bottom line here is that the world of mental images, concepts, information, and knowledge is where you will find the best fit, and the most potent applications, of convergence technologies. I tell my students that anything that you can "get into your head" can be digitally represented if it's not already, providing additional fuel to what is becoming a raging convergence fire. Parts III and IV will help you to better understand how, from a technology standpoint, this is happening.

In the next chapter, we'll profile some of the challenges that many large organizations face as convergence technologies continue to empower individual customers and employees.

4. Corporate Challenges

Large corporations have been the primary beneficiaries of technological innovation in prior technology eras. The mainframe enabled corporations to more effectively manage increased business volume. And while the PC revolution provided consumers with an opportunity to creatively leverage computer technology, it was a boon to large corporations once enterprise networks linked the desktops throughout the global enterprise. Corporate technology deployments resulted in improved corporate communication; more efficient and effective business processes and production mechanisms; and ultimately, products that would provide customers with cost savings, better service, and greater choice and quality. Now, in the age of digital convergence, the pendulum has begun to swing in favor of individual creator-collaborators who are empowered by new technologies (see Figure 1).

Figure 1

Prior Technological Eras *Rise of the corporation*	**The Convergence Era** *Rise of the creator-collaborator*
· Scalable technologies enable businesses to cost-effectively support a large set of product offerings and a very large customer base.	· Corporations struggle to implement and support complex convergence technologies enterprise-wide.
· Corporations stockpile knowledge resources, technology, and expertise.	· Knowledge and expertise are readily accessible via the inexpensive and readily available technologies of convergence.
· Talented individuals seek employment with large corporations that offer attractive salary/benefit structures.	· Many creator-collaborators possess the talents and initiative to profit handsomely outside the framework of a large company.
· Corporations determine where employees work and live, and where, when, and for how long they must work.	· Many creator-collaborators are gaining control over where and when they work.

While corporations will benefit from new convergence technologies, especially those in impalpable industries, larger organizations are often blinded to the power of individual technology-enabled knowledge workers. These employees should be free to aggressively explore and use new technologies. I believe that the most useful corporate applications of convergence technologies will bubble up from trenches.

Don't get me wrong. There are still significant opportunities for corporations to benefit from the use of new technologies. These applications will most often enable the enterprise to realize internal efficiencies and reach out beyond the boundaries of the corporation:

- improve interactivity with global business partners
- increase supply-chain efficiencies
- enhance customer service
- enhance employee productivity
- increase "touch points" for consumers
- access new global markets
- improve employee access to information

These types of initiatives have been the focus of many enterprise implementations. Unfortunately, corporate leaders tend to view technological opportunity purely from an aggregate and generic corporate standpoint. This perspective masks the ability of the technology to empower individuals. Many corporate leaders also seek to maintain a firm control over technology, and actively resist placing the full power of convergence technology in the hands of individual employees.

The next two sections briefly outline these two messy aspects of corporate technology use, which are often swept under the carpet by corporate IT types and by the technology and business media.

Enterprise Systems Challenges

For decades corporate technology decision-makers have sought to implement sophisticated and expensive information systems to enhance internal productivity, communication, information management, and decision-making across the enterprise. Back in the early 1980s (when I got my start as an IBM mainframe programmer), corporate IT was touting the benefits of automation and database management with a vision for enhanced operation of the corporate machine; Fourth-Generation Languages promised to make it easier to create and manage mainframe-based enterprise systems while also easing end-user data access; and EDI (Electronic Data Interchange) was being implemented for information sharing among companies. Since then, the technical boundaries have become blurred, and potential corporate impact may be greater than what we expected back in those days, but the underlying motivations, and many of the underlying concepts, are the same.

Often, the promise of complex, enterprise-wide technology implementations is never fully realized. Over the past few years, expensive client/server systems from SAP, i2 Technologies, Oracle, and others have struggled to deliver the expected return on investment.

- In June of 2000, Nike began implementation of a $400 million supply-chain system, based on i2 software. During implementation it found a major glitch that resulted in inventory shortages and a third-quarter 2001 sales reduction of $100 million.*

- Agilent Technologies blamed a botched Enterprise Resource Planning (ERP) implementation, based on Oracle software, for a third-quarter 2002 revenue shortfall of $105 million, with a $70 million shortfall in operating profits. CEO Adrian Dillon com-

---

* Tim Wilson, "Supply Chain Debacle," *InternetWeek*, March 1, 2001 (www.internetweek.com/story/showArticle.jhtml?articleID=6401505).

mented that this implementation experience was "about normal but not significantly better than the average experience."*

- During a seminar with a group of middle managers, I opened a discussion on enterprise-wide software solutions. ERP software quickly came up, and one of the students shared with the class a standing joke at his firm: SAP (one of the top providers of ERP software) stands for "Stop All Production."

Two of the next big things on the corporate IT agenda are portals and Web services. A portal is a Web-based framework that allows customized access to Web-based applications and information. Web services are components of a new architecture (discussed in more detail in chapter 12) that enables distributed network access to information and business functions. As central IT works to deploy these complex technologies across the enterprise, I fully expect history to repeat itself as expensive implementations initially fail to meet over-hyped expectations. Why? Because of a few harsh realities which together constitute the dark side of enterprise information systems deployments.

Technical complexity

Software and systems deployed throughout a large enterprise are complicated and expensive to implement. In the mainframe era, the drill involved implementing applications software, properly sizing and administering the mainframe, and maintaining a low-speed terminal network and a set of CRT terminal stations. While this was not simple, it was cake compared to today's drill.

* Tom Smith, "Price of Latest Software Failure: Well Over $100 Million," *InternetWeek*, August 20, 2002 (http://www.internetweek.com/story/showArticle.jhtml?articleID=6400127).

Enterprise systems now require an army of technical support staff members who must oversee:

- server and client applications software
- database, Web, and application servers
- switches, routers, and PC operating systems
- security mechanisms, user directories, and other utilities

Potential points of failure are many. The ability of technical staff to estimate loads and to size the infrastructure accordingly is limited. Unfortunately, much of this is still art, not science, and problems are not at all uncommon.

Functional variation and the common denominator problem

Enterprise systems are developed by programmers and analysts who try to understand the functionality needs of the target user-base. This is manageable for a small number of users. But when the number of users increases, the ability of the application to respond to the variety and complexity of user needs and expectations diminishes. An associate dean in the Whitman School of Management effectively captured this downside for me when he described why the university's enterprise-wide Peoplesoft system is so frustrating. He said that the developers design the system for some mythical generic user, and in the process "just piss everyone off."

Exception overload

Any information system always has to deal with exceptions: user activities that don't fit the design; transactions that are irregular; users who are irregular; information that is incomplete; and on and on. If a system is small, the exceptions are rare and can usually be dealt with rather easily. However, as a system scales up, the number and diversity of exceptions increases. This poses a considerable burden on either the system or its users. I call this *exception overload.*

System managers often deal with exception overload with policies or procedures that are limiting or awkward – which usually results in a poor user experience for those individuals who generate the exceptions. Users are left to accept a system that doesn't really meet their needs, or implement awkward, sometimes costly workarounds.

The realities described above will continue to impede complex internal and external enterprise systems that are scaled to serve many users.

Constrained Creator-Collaborators

In many organizations, restrictive, centralized management and control of technology constrains creator-collaborators.

Personal Computers That Aren't So Personal

If you're like most people, the desktop computer that you purchased for your own personal use came with a hard drive that had a pre-installed operating system such as Windows 98, Windows ME, or Windows XP (depending on how old your PC is).* Most PCs come bundled with applications programs as well, such as Microsoft Office, image editing software, security software, and so on. The PC user has the option of installing any software or connecting any hardware – scanner, printer, camera, etc. – that he or she chooses. The operating systems and software that can be preloaded on new PCs vary considerably, enabling PC

* Yes, I know some of you love your Macintoshes, but I maintain that a PC-centric view of convergence is fair, given the market dominance of the PC. After an aggressive advertising campaign in 2002–2003 Apple's share of the worldwide PC market remains well under five percent. From Jon Fortt, "Few Customers Switching from Windows," *Mercury News*, July 17, 2003 (http://www.siliconvalley.com/mld/siliconvalley/news/local/6324496.htm).

buyers to customize systems based on unique personal interests and needs.

In a large organization, centralized technical and user support for multiple operating systems and diverse application software is difficult and expensive. For example, some operating system configurations or unusual applications can cause pesky bugs or conflict with enterprise application software. In a large corporation these quirks can add up.

Total cost of ownership (TCO) is a concept that captures this overall cost of desktop support in a large enterprise. IT managers work very hard to reduce TCO, and this often revolves around a set of software control and standardization strategies.

- Corporate IT managers may negotiate volume purchase agreements with PC manufacturers that specify that all systems will ship with a standard operating system and software.
- Some desktop system managers may totally replace the operating system and software that ship on the hard drive with a uniform operating system and applications combination. This can include software developed by the corporation in addition to standard productivity software. This standard software configuration is called an *image,* and there are a variety of software utilities on the market for capturing an image from a model computer and replicating it on target machines.
- Corporate computers may be fitted with what are considered business-class operating systems – such as Windows NT, Windows 2000, and Windows XP Professional. These operating systems allow the central IT group to disable the ability of end users to install new software. These are called "locked-down" desktops.
- CIOs may put in place a policy that restricts technical support to a small set of authorized software programs, or a policy that de-

nies all support for any user who has installed non-standard software on his or her system.

While there are some good reasons for these approaches, I see a significant problem as well. Standardized, locked-down PCs are most often deployed on the desks of knowledge workers, including many who are creator-collaborators. The PC is an incredibly powerful device, and many employees want the ability to customize their desktops to improve performance and productivity. And many possess the creativity and technical know-how to develop valuable workplace innovations.

I work with university undergraduates who are tomorrow's knowledge workers, and I can assure you that a growing percentage consist of young, tech-savvy creator-collaborators. Many of the best and brightest will reject corporate technology constraints – and either flee the enterprise or become in-house technology renegades. To retain these employees and leverage their abilities, a corporation must seek to empower its technologically adept human resource with new convergence technologies. Clearly, savvy corporations must revisit the TCO agenda, as more and more creator-collaborators join the workforce.

Batten Down the Hatches

Corporate IT managers are concerned not just with operating systems and software, but with what the employees actually do with their PCs as well. Many who work leverage the PC to its fullest potential are greeted with a decidedly bad attitude when confronting the central IT group. Typically, IT concerns are related to security, bandwidth, and productivity.

Security

Viruses, worms, hackers, and other security intrusions are a threat. To combat this, the IT department usually sets up a firewall that constrains Internet access from the outside world. The IT department may also:

- strip email attachments from incoming email messages
- force employees to use complicated passwords, and to frequently change their passwords
- set restrictive policies on Web surfing
- restrict the deployment of departmental and workgroup servers and networks

While there is a real need to address security concerns, the typical restrictive, intrusive IT response means employees are often unable to effectively share information, collaborate, and communicate with others. At the Whitman School we often encounter this when interacting with executive students working behind corporate firewalls; with corporate users of our live Webcast and on-demand streaming video content; and with employees and students who become so frustrated with on-campus security hurdles that they actively pursue ways to circumvent the perceived security burdens while vocalizing their displeasure.

In many corporate environments the constraints placed on employee use of technology – in an effort to ensure a secure use of computing and information resources – are considerable and may need to be re-evaluated. While this is clearly not a popular viewpoint within the IT community, the cost – in terms of disabled communication and lost productivity – is also too often ignored by the corporate IT bureaucracy. In the most severe cases new convergence technologies are viewed almost entirely as a threat rather than as an enabler.

A study of analysts in the CIA's Directorate of Intelligence* concluded that an almost exclusive focus on new technology as a threat rather than enabler ultimately resulted in a diminished capacity for CIA analysts to use technology effectively. According to Bruce Berkowitz,

* Bruce Berkowitz, "Failing to Keep Up With the Information Revolution," May 2003 (http://www.cia.gov/csi/studies/vol47no1/article07.html).

the author of the report and a fellow at the Hoover Institution think tank at Stanford University, this "technology gap" constrained the ability of the CIA to respond to the war on terrorism following the September 11, 2001, terrorist attacks. He summarized the long-term implications of the constraints imposed on CIA analysts as follows:

> "By not encouraging analysts to use information technology more creatively, the [CIA] is hurting its future. Most 'killer apps' – unusually powerful and effective applications software developments – originate from users, not programmers. The first step in developing software is recognizing that a need exists, and users are the ones at the scene who know what they need."

While the security threats that most large organizations see may differ from what the CIA may encounter, the need for workers to use technology in creative ways will continue to be crucial to successful organizations of all types and sizes.

During a conference I attended recently I heard an IBM security expert address this issue. He suggested that most organizations react to potential security threats by focusing almost exclusively on prevention – which often results in restrictive technology use for end users. He suggested that a balanced security response that does not unduly constrain use of new technology can be achieved with a reduced focus on prevention and a greater focus on intrusion *detection* and *response*.

I am not a security expert and I don't pretend to have any easy answers. But I hope that IT industry and IT departments can begin to offer solutions that do not restrict or constrain the power and flexibility of convergence technology. Organizations that actively pursue this approach stand to gain a competitive advantage in the long run.

Bandwidth

A portion of the information transmitted across a corporate Intranet is deemed mission-critical, usually crucial operational information and financial transactions. To prevent bottlenecks that could interfere with mission-critical traffic and to save money on network provisioning and operation, the IT department may give a lower priority to transmissions it considers less important, or it may block those transmissions altogether. For example, some companies disable access to external Internet streaming media content, except for live or on-demand streams approved by a corporate official. A company may also ban bandwidth-hungry rich media applications, such as peer-to-peer file sharing programs, media players, and conferencing tools such as Microsoft NetMeeting.

The use of rich media technologies for professional communication and information sharing is growing rapidly in phase 2. At the same time, the cost of increasing the capacity of network plumbing continues to decline. For many organizations, constraining employee access to save on bandwidth will ultimately prove to be a shortsighted misstep.

Productivity

I heard on a drive-time radio show that unauthorized employee Internet access is the number one reason cited for employee termination. I don't know if this is true, but clearly, companies see many Internet activities – Web surfing, chatting, email, instant messaging – as a significant threat to productivity. Corporate IT departments have the ability to monitor employee Web access and email, and in many large organizations this is the preferred response to the potential for non–work-related use of technology. I believe it is an inappropriate response. Successful companies in the age of convergence will not monitor and micromanage Internet use, but will instead choose to manage employee activity based on larger goals, objectives, and deliverables; they will encourage the individual exploration of new technologies, ultimately gaining a competitive advantage over competitors who ban, block, or stifle technology use.

The University Offers an Alternative Model

The university has become a growing, thriving virtual learning community. The wide range of convergence-based human interactions within this community offers a preview for the future of business where there is:

- near ubiquitous broadband
- a diverse community of adept technology users
- global interactivity with and among a variety of external constituents

Corporations that expect to thrive in the age of convergence will need to empower employees and facilitate the technology-based human interactions now evident in today's university learning community. At the university, and in the successful 21st-century enterprise, it's really all about people – how they interact, communicate, collaborate, and share information.

The university community includes four types of individuals:

1. *Staff:* Most staff members are knowledge workers, and a vast majority work with computers extensively.
2. *Faculty:* Faculty members are knowledge creators, and many possess strong technical skills. A significant percentage of faculty are also technologically self-sufficient, since they regularly secure significant funding for technology through grants and other external sources.
3. *Students:* A typical university student owns his or her own computer and has a broadband connection through a wired dorm, DSL connection, or cable modem (usually shared with other students in a residence). Many students have notebook computers

with wireless network capability. The university has very little control over student desktop hardware and software.

4. *External constituents:* Technology-based interactions with outside constituents – alumni, business professionals, parents, colleagues and distance learning students – extend the university well beyond its physical boundaries.

Support for the university's diverse and fiercely independent users offers what I think is a good role model for fostering innovative and effective technology use in a large organization. Corporations are increasingly driven by ideas and knowledge – and many will eventually see the value of the university's technology support model.

- The university strives to support diverse applications of technology across the wide range of disciplines, departments, workgroups, and individuals within the university community.
- The university supports the creative use of digital technologies outside the scope of enterprise business processes. Technology offers great long-term potential in some application realms that are not easily tied to short-term, bottom-line business outcomes.
- The university works hard to coordinate technical expertise across the institution. The university has a central IT group, but many technical support personnel work out in the trenches where their reporting lines are to department heads, not central IT managers.
- The university computing environment is more open and customizable than the locked-down corporate environment. A very conscious effort is made to balance the need for security and reliability with the need to share information and ideas across the campus, and throughout the world.

People: the Heart of the Corporation

To end this chapter I share with you a brief anecdote from one of my executive classes that underscores what I believe is a crucial notion: that convergence technologies will serve humanity most effectively in a granular, individual context.

Customer Relationship Management is a fairly popular technology-based methodology for systematizing human interactions. CRM technology gathers and integrates detailed information about customers, and makes that information available to customer service personnel and related systems and departments – enabling the company to offer personalized services. The assumption is that this will create good will and produce happy, loyal customers. This sounds good in theory, but there are some potential drawbacks.

A student in one of my classes was employed by a major hotel chain. She did a project on customer relationship management, and in her presentation, she described a hotel system that captures data on the needs and habits of the high-paying repeat customers. If one of these customers orders extra towels or requires a refill of Diet Pepsi in the minibar, it gets entered into the CRM database. Hotel employees are instructed to access this information to try and anticipate and accommodate unique customer needs and desires. After her presentation, the class discussion quickly focused on how hotel chain employees would record and share this sort of detailed information. The students clearly perceived a "Big Brother" scenario, and student reaction ranged from concern to outrage.

I was struck by the desire of the large hotel chain to essentially replicate the personal touch one might receive from a well-run mom-and-pop hotel. To me, it seemed like CRM technology was being used to try and give the corporation a human touch – a heart, if you will. Instead, the enterprise-level deployment smacked of a heartless Big Brother.

Ultimately, the use of new technology in any organization of any size comes down to the individual human beings – employees and customers – who have unique needs and agendas. Even the largest multinational corporation is essentially a very large group of individual human beings working together to achieve a common goal. These individuals are the heart of the corporation, and the power of convergence technology will ultimately have the greatest impact in their hands.

5. THE WINNERS

Over the last four chapters, I've presented my view of the issues and realities that are determining where the most significant impact will occur as digital convergence technologies take hold. That impact will continue to be felt across our society and our economy, throughout our institutions and corporations, and within our own individual homes and places of work.

Here I want to bring this all together with my take on the biggest winners in the age of digital convergence – and in the process reinforce the opportunity that digital convergence offers for the creators and collaborators among you who want to attack the individual opportunities that convergence presents.

I know I'm going out on a limb here, and many of you may disagree with my assessment. That is just fine – in fact it is welcome. As I share with you my winners, I want you to think through your own list of winners (and maybe losers too). I use this technique in my classes and seminars as a means of focusing on where the greatest impact will be as the digital convergence era unfolds. I tell students to define "winner" any way they choose, and to be unconstrained in their thinking. While I suggest a set of winners below, your list may include some very different types of winners – perhaps a particular industry, or a type of profession, or maybe the companies that do business in a particular part of the world. This is an exercise that anyone can do and which serves as a good first step toward linking the opportunities and potential outcomes of this era to your own personal and professional perspective, ambitions, and objectives.

So, before you turn the page to check out my top five, I urge you to set down the book for a few minutes and reflect on this: from your own unique standpoint, what are the most important factors that you see shaping the future of digital convergence, and who will be the "big winners" in your mind?

Before I roll out my top five, I first want to summarize for you what I believe are the five keys to the second phase of the digital convergence phenomenon:

1. *The Golden Rule of Technology:* the price of any new technology will decline as its functionality, power, and ease of use will increase (this important rule was introduced in the first book)
2. *Broadband:* the need for speed; we need to see high-speed consumer access widely deployed before we see the digital convergence phenomenon successfully move to the next phase
3. *Rich media:* audio, video, and other rich media will become integral to our Internet interactions
4. *Impalpables:* conceptual goods and services and individual impalpables make up the "oxygen" that is delivered through Internet "arteries," bringing life to our thriving, global "organism" of activity and opportunity
5. *Special interests:* digital convergence technologies really shine when used to support an ever-increasing array of special interests (the technologies of convergence are very useful in support of individual, group, and community-of-interest interactions).

Note that I did not include mobile Internet. Yes, this is an important development, but what it enables is, in my mind, largely *convenience* – which is just not as important as some other emerging convergence technologies that are more closely aligned with the four drivers outlined above. I'll develop this further in the final section of the book, where I focus on the use of new convergence technologies and applications, including mobile Internet.

The Covell Top 5

Big Winners in the Age of Digital Convergence

(drumroll, please . . .)

#5 – Mass Market Innovators

Many mass market dot.coms have already gone belly-up. But while mass market e-commerce player Pets.com was going bust and Amazon.com was still promising to turn a profit, eBay was becoming a rock-solid enterprise and the original Napster was becoming an overnight slam-dunk like I haven't seen since the Winnebago, even though it never did turn a profit.

I believe that the mass market innovators who stand the greatest chance of success are those that do one or more of the following:

- focus on impalpable products (e.g., Travelocity)
- sell digital convergence technologies that have mass appeal (e.g., PCMall.com)
- provide valuable services and infrastructure for creator-collaborators and consumers (e.g., eBay)
- build a broad-based community (e.g., Amazon.com)

There will be some success stories, but, hey, I'm not about to sink my 401(k) into any Internet mass market innovator anytime soon. The fracturing of markets and leveling of playing fields that convergence technologies enable is real. That's the biggest single reason that mass market innovators appear on this list at number 5 – dead last.

#4 – Consumers

iPod, Xbox, XM Radio, TiVo, and many other new convergence-based consumer gadgets are great. Merry Christmas and Happy Hanukkah!

As I mentioned in chapter 3, consumers will continue to see more choices and more products. And they'll have more information to draw on when making purchase decisions.

Still, passive consumers will remain shackled by the technologies and the constraints imposed by mass market providers who seek to deliver only the most broadly useful (and lowest common denominator) technologies and interfaces. Thus, consumers come in at number 4.

#3 – Business-to-Business Vertical Market Innovators

Vertical markets are all about special interests, which is the bread and butter of the human interaction that convergence technologies enable. Enterprises focused on vertical markets are good candidates for innovative B2B services and resources. Entrepreneurs in a variety of specific industries will successfully pursue opportunities based on new mechanisms for individuals and organizations to interact and collaborate. B2B vertical market innovators come in at a solid number 3.

#2 – Technology Providers

Digital convergence promises to continually give rise to more effective human and organizational interaction and information sharing over the course of the next few decades. Businesses, governments, and not-for-profit organizations, along with individual creator-collaborators and consumers, will all continue to pursue emerging technologies that can deliver on this promise. Companies that provide the needed technologies and technical infrastructure will thrive well into the future. Remember, we are just entering the second phase of this phenomenon – and although the early part of this century saw a tech meltdown of sorts, the tech sector is beginning to come back, and I believe it will soar once again.

Yes, we will continue to see missteps, consolidations, and failures. But in general, the hardware providers, networking providers, software companies, telecommunications providers, Internet Service Providers, and a variety of other technology infrastructure providers will flourish. Some of you may argue that technology providers are the real number 1. I disagree, but I do concede them a clear number 2.

#1 – Creator-collaborators

Creator-collaborators, unlike passive consumers, are unshackled in their use of new technology and will be able to profit both personally and professionally. For those of us who work to leverage the technologies of digital convergence in exciting and productive ways, the opportunities will be endless.

I predict that the ranks of the accomplished creator-collaborators will swell over the course of the next decade. And I believe that within the next decade we will see widespread recognition that we have indeed changed the world.

Part III

The Forces of Technology Evolution

In the first book I surveyed technology developments in the underlying realms of computing, communications, and digital media. As we move into phase 2, technological innovation and convergence in these realms continues to propel the evolution of digital convergence. Here is a quick sampling of recent developments:

Digital media. *Quality and functionality continues to improve. We have better displays; higher quality and less expensive digital imaging and video technologies; better, cheaper color printers; and more powerful media editing tools. We also find new ways to move large amounts of media between devices as high-speed transfer technologies (such as USB 2.0 and FireWire) and high-capacity writeable and removable media technologies (particularly CD and DVD burners) are being built into many more devices.*

Communications technology. *We continue to see developments in optical networking including faster backbones based on dense wave division multiplexing and optical switching (which eliminates bandwidth-constraining conversions between optical and electrical signals). We have also seen a growing uptake of broadband technologies in the last mile, and we're beginning to see more Wi-Fi networks and third-generation wireless cellular technologies.*

Computing. *Moore's Law continues, and – with promising developments in fields such as nanotechnology and photonics – should persist into the foreseeable future. Computer hard drive and memory in-*

novations enable greater capacity at increasingly reasonable price points, and developments in other computing components will continue to provide steady price/performance improvements. On the miniaturization front we're seeing an uptick in the number of increasingly powerful portable, ultra-portable (handheld), and embedded computer systems. This trend will continue to evolve over time, and many believe that at some point the dominant computer – which I define simply as the general-purpose computer people most often interact with – will indeed be integrated into the fabric of our lives – in our entertainment appliances, our pockets, our cars, and maybe even in our clothes.

Since I wrote the first book, the telecommunications, media, and technology industries have all begun to more aggressively pursue what is believed to be the powerful commercial potential of convergence technologies. While technical innovation does indeed create opportunities for creator-collaborators, other factors – such as technical standards, marketplace competition, and government intervention – have a growing impact on innovation; and these factors frequently enable or constrain the evolution, manufacture, and deployment of new convergence technologies. A phase-2 creator-collaborator needs a more comprehensive understanding of the intersection of the technical and nontechnical factors that impact convergence if she wants to better understand technology evolution and more easily and effectively identify opportunities to deploy new technologies. This section explores this intersection.

This section also introduces a growing tension: As the convergence phenomenon rolls on to provide an ever-richer range of functionality in an increasing scope of deployments, creator-collaborators are challenging corporate control over markets and technology. In this section, I'll attempt to characterize this mounting tension as it relates to the evolution and adoption of convergence technologies. This tension is also bleeding into other realms, such as entertainment – and we'll take that up in the next section.

6. TECHNOLOGY LAYERING

You push the "on" button on your Windows PC to power it up (I know, some of you have a Macintosh or Linux system instead, but most of you use a PC). What typically happens next – what tech-heads call "booting up" – is really very instructive.*

1. First, you probably see some sort of "on" light, indicating that the unit is powered up.
2. Then you see a black screen with some information about your hardware configuration.
3. Next, Windows begins to load. You typically see a line or two indicating the version of the operating system that will load; then you see a screen with a logo indicating that Windows is loading. After several seconds (sometimes *many* more seconds on slower machines) you see your desktop wallpaper and some icons.
4. As Windows finishes loading, you'll see graphical displays or messages as startup programs slowly load into memory and then appear as miniature icons in your system tray (a small panel in the lower right corner of your screen).
5. Once your system has finished booting, you double-click an icon or go to the Start menu to fire up an application program to perform some specific task. The software you select loads into memory and presents you with its interface. Now you're now ready to do some work, or maybe have some fun!

---

* Macintosh computers and PCs running the Linux operating system have similar boot sequences.

This simple sequence of displays roughly illustrates the technology layering that underpins the desktop computer system that you and I know and love (OK, occasionally we hate it too!). And it offers an intuitive opportunity to introduce one of the most important technological principles of the digital revolution: the implementation – and the evolution – of digital technologies as a series of interdependent layers. All convergence technologies are architected with this same layering logic.

While technology layers are interdependent (no layer can work without other layers) any particular layer is independent of the layers that are not immediately above or below it. This isolation enables technologies to evolve fairly quickly. For example, PC hard-drive technology, which is at a low-level layer, can evolve without regard to software applications, which are at a much higher layer.

This sounds all good in theory, but how does layering *really* work? I get that question from students regularly. Let's examine the layers of a Windows PC in a little more detail to flesh out the concept. For a somewhat simplified view of technology layering, I'll place the technology into three broad categories: physical layers, intermediate layers, and application layers (see Figure 1).

Figure 1. PC Layering

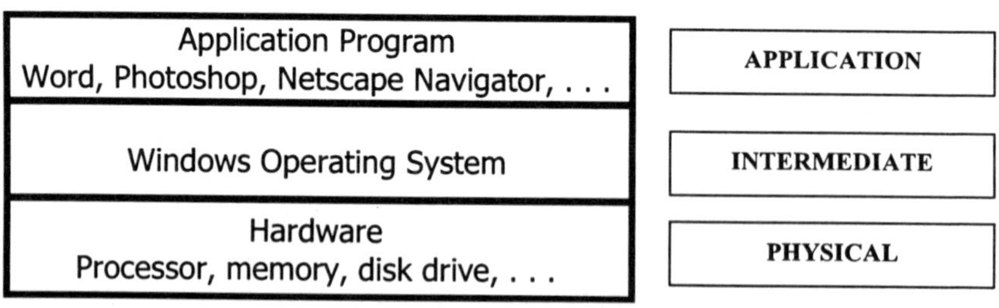

Physical Layers

At the lowest layers, the PC relies on the physical properties of matter, electricity, magnetism, lasers, and optics. Digital ones and zeroes – the basis of all digital media and software – are stored in magnetic fields on spinning disk drives, in voltage signals in Random Access Memory, and in the reflective surface of CDs and DVDs. Processing of digital instructions and information occurs in dynamically altered electronic circuits on silicon chips.

Scientists and engineers have worked out standardized mechanisms for input, output, and transmission of electrical, optical, and magnetic signals carrying digital instructions and information. Much of this is coordinated in a single unit, called a motherboard, which can be mass-produced by hardware providers. Physical layer standardization also enables production of compatible components – such as hard drives, video cards, and memory modules – by competing hardware providers. This competition stimulates rapid evolution of PC capacity and performance.

Computer hardware and software engineers develop computer programs specifically designed to control hardware components. These are called device drivers. They constitute an interface between the physical and intermediate layers, enabling software program control of hardware devices. Some of the engineers who develop device drivers work for companies who provide PC hardware components, such as Creative Labs (sound cards), Hewlett-Packard (printers), Western Digital (disk drives), and Sony (CD-ROM drives). Others work for Microsoft, the creator of the Windows operating system.

Intermediate Layers

Windows, which is a set of layered software programs collectively referred to as an *operating system,* resides in what I call the intermediate layers, directly above the physical layers.

An operating system does three very important things:

1. It has a user interface. The user interface enables users to load and execute application programs and interact with the system – managing files, changing system settings, adjusting speaker volume, and so on. GUI

2. It manages the hardware. For example, when you copy a file by "dragging" it from one folder to another you can hear the disk spinning while it reads and writes the digital ones and zeroes that comprise the source file. The operating system commands the hardware to perform the copy through calls to the appropriate device drivers.

3. It has a programming interface for application developers. Programmers use a programming language, such as C or C++, to create application software that has the processor logic required to perform specific functions. Application programs include commands that are submitted to the operating system's Application Programming Interface (API), which translates those commands into action. This can be sending data through a network connection, writing data to a file, or displaying an image on the monitor.

Powerful Abstractions

One of the keys to the layering paradigm is the introduction of abstractions that provide simple, straightforward access to the functions of lower-level layers. These abstractions are clearly evident in the intermediate layers.

- The Windows operating system has a graphical user interface (GUI) that provides visual cues indicating what is happening in the underlying hardware. On the Windows "desktop" you may see an icon which represents a file. If you open the "My Computer" icon, and drag the file icon onto the icon that represents your floppy disk drive (drive A:\), the operating system will instruct the hardware to copy the bits corresponding to the file from the magnetic surface of your hard drive to the magnetic surface of a floppy disk. As this occurs you may see a graphical image of a paper moving from one folder to another. These are all abstract, graphical representations of what is going on in lower-level layers.

- Software programmers create text-based programs that include program statements that "call" underlying functions – via the Application Programming Interface – using symbolic text parameters, or variables, that specify the details of the functions to be performed. The API makes it relatively easy for programmers to create useful capabilities and interactions that combine program logic with the manipulation of underlying hardware and data objects.

Abstractions such as these, and the black-box functionality they provide, form the basis of layering – as each higher-level layer adds a new layer of abstraction.

Application Layers

PC application developers create software programs that provide useful functions and capabilities for end users. Here you find the application software (or simply applications) that you use to accomplish various tasks – such as writing a letter, composing an email, accessing a corporate database, editing a video, crunching numbers, or viewing streaming media. I place these at the highest set of layers, which I call the application layers. In the application layers you also find the greatest degree of abstraction, enabling friendly, intuitive user interfaces and functions.

Since I'm sure you have experienced some technologies that reside in the application layers, this is a good place to explain two aspects of technology layering that can cause some confusion.

1. *Multiple layers:* I always refer to physical, intermediate, and application *layers* in the plural. Some of you may wonder why.

Aren't all PC software applications in a single application layer? Not necessarily – there are layers within the layers.

2. *Gray Areas:* Sometimes, a specific technology has multiple uses that could logically place it in more than one set of layers.

Take Excel, for example. You and I use Excel to manage and analyze information, and it seems clear that Excel is an end-user application. Yet Excel can also be used to create forms-based interfaces for entering data and displaying predefined reports and graphs. You can even hide much of the normal Excel interface, making a programmed spreadsheet look and feel like a customized application. These Excel-based applications can comprise a new layer that sits "on top" of the Excel layer. So, while Excel is an end-user tool that clearly goes in the application layers, it is also a programming environment. And I usually place programming tools in the intermediate layers.

You can begin to see the potential for multiple application layers and technologies that fall into a gray area between intermediate and application layers. There are also multiple layers in the physical and intermediate levels, and gray areas between intermediate and physical layers.

As an example of layers within layers, Figure 2 offers a much more detailed view of the layering within the Windows operating system (here you can see up to eight layers within the intermediate layer operating system – and this is still a somewhat simplified view). Clearly, layering can get pretty complicated.

Please don't get hung up on these little twists. I've offered a simplified view of layering to help you grasp the basic notion of technology layering so you can begin to get a sense for how it fosters technology innovations. Being able to simplify and identify key technology layers is very useful when trying to understand the evolution of new technologies – in a wide variety of technological contexts. Over the course of the next couple of chapters we will explore this in more detail.

Figure 2*

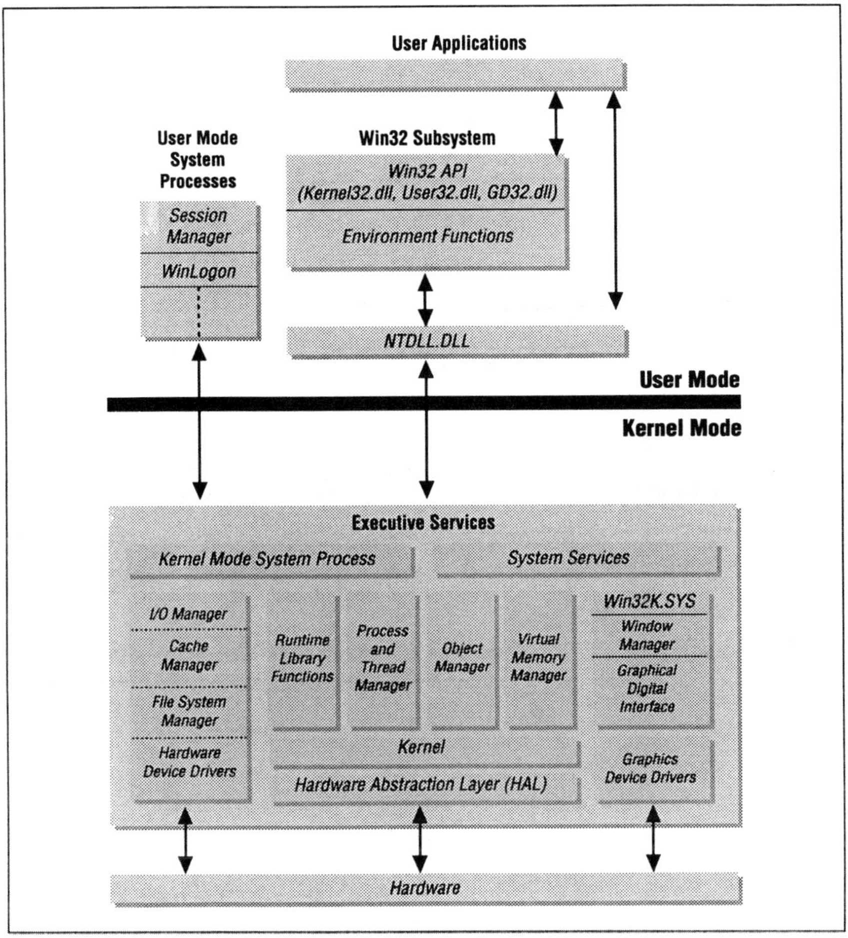

* From an excerpt from Steven Roman and Ron Petrusha (Editor), *Win32 API Programming*. © 1999 O'Reilly and Associates. (http://www.microsoft.com/technet/treeview/default.asp?url=/technet/prodtechnol/ntwrkstn/evaluate/featfunc/winarch.asp).

Beyond the PC

Layering is an approach to technology that applies to all digital convergence technology realms. Digital networks are implemented in layers, as are Web technologies and digital media technologies. To reinforce the broad applicability of the layering framework, let's look at another example of technological layering: a digital satellite television service such Dish Network or DirecTV (Figure 3).

Figure 3. Digital Satellite Television Layers

Application	· One-way, one-to-many broadcast of selected entertainment and information programming*
Intermediate	· Digital video compression algorithms.
	· Video format specifications – including picture dimensions, frame rate, and interlacing scheme
	· Protocols and format standards for the digital video signal
Physical	· Technologies that enable the use of electromagnetic spectrum to transfer digital information from satellite to mini-dish satellite receiver
	· Electronics and cabling used to convert and then transmit electronic signals from dish to destination television
	· Electron beams that energize phosphor elements on a cathode ray tube (CRT TV monitor) to generate a video image

With this example, you can see that technology layering works in digital convergence contexts beyond the PC.

* Like cable providers, satellite providers are moving to broaden application offerings. We'll revisit this in upcoming chapters.

You may also notice that the applications mix is much more constrained when it comes to satellite television. Legacy technology realms, such as television and telephony, typically lack the flexibility – i.e., flexible interactions and multiplicity of applications – inherent in the context of the PC, the Internet, and the Web. We'll come back to this issue in later chapters.

In the next chapter we will begin to explore the importance of standardization as it relates to layered technologies.

7. Technology Layers and Standards

There are many convergence realms where innovation and adoption of new technology is dependent on the creation and voluntary industry support of layered technology standards. In this chapter we'll review a series of examples to help further your understanding of technology layering and the role of standards.

The IBM Mainframe: Proprietary Standard Technology

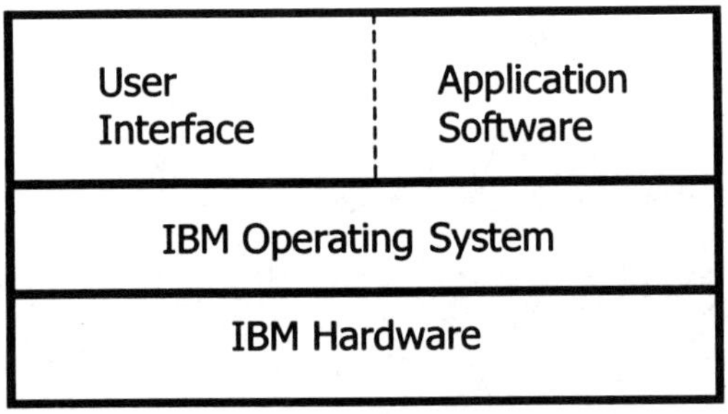

I first encountered standard information technology layers when I worked on an IBM mainframe in the 1980s. While there were so-called "plug compatible" mainframes from competitors such as Amdahl and Control Data Corporation, the IBM mainframe was the dominant technology of the time. Its hardware and operating systems were physical- and intermediate-layer standards – in this case, proprietary standards, since all the hardware and operating system technologies were owned by IBM. Competitors had to reverse-engineer IBM's technology in order to create compatible alternatives.

Software AG, Computer Associates, and other software providers sold database software and applications for IBM mainframes. But much of the software in the intermediate layers, such as database management systems and programming languages, were also provided by IBM. And IBM offered a variety of application software as well.

There were some positive aspects of IBM mainframe dominance. For example, IBM mainframe knowledge and expertise was a ticket to gainful employment, and a substantial pool of talent emerged to feed the growing mainframe-based corporate engine. Another plus was the portability of data. As a mainframe programmer, I spent many hours writing or reading data on IBM mainframe magnetic tapes shared with other organizations.

There were also some downsides to the IBM standard. Chief among them was a lack of competition. IBM owned the standard and they owned the market. Yes, there were a handful of plug compatible mainframe alternatives, and minicomputers were a viable alternative for smaller organizations. But for most mainframe-era data processing managers, anything with IBM on it was seen as a safe choice. With a lack of competition, Big Blue was able to keep mainframe prices high and focus on those innovations that would further extend its grip on corporate IT budgets.

I was so pleased when we began to see the rise of network computing and a proliferation of minicomputer, server, and workstation computers – from DEC, Sun Microsystems, Hewlett-Packard, Prime Computers, and others – with enough computing power to begin to challenge the dominance of the IBM mainframe.

The IBM Compatible PC: Shared Technology Standards

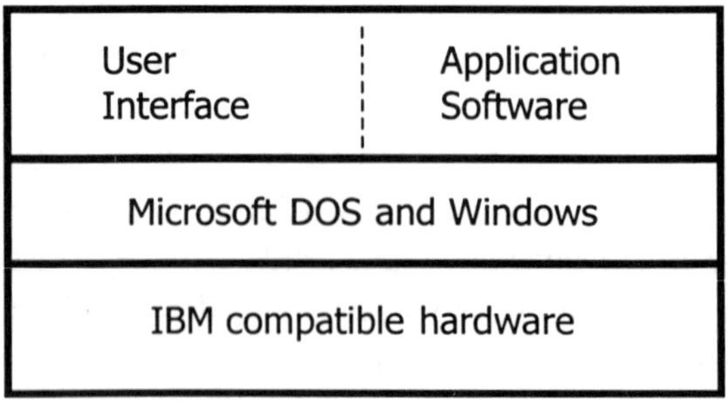

When the IBM PC was introduced back in the 1980s, an important step toward intermediate layer standard technology was taken when an agreement between IBM and Microsoft allowed IBM to use the MS-DOS operating system, but also allowed Microsoft to license the operating system to any other hardware manufacturer. Several computer vendors were able to reverse-engineer a core component of the IBM PC, called the BIOS (Basic Input Output System), obtain a license to run MS-DOS, and then create competing computers, which were known as IBM-compatibles, or clones. Many hardware companies were eventually able to manufacture PCs that would run MS-DOS and all of its application programs. This ultimately resulted in a huge marketplace for IBM-compatible PCs – with Compaq, Northgate, Gateway, Micron, Dell, and many other new and existing hardware manufacturers offering competing PCs.

MS-DOS rather quickly became an operating system standard, and although it was proprietary, it was available to any hardware vendor. This portability of MS-DOS across alternative hardware platforms is known as *interoperability,* a powerful notion that is at the front and center of the standards movement.

With today's personal computer, the same sort of proprietary standardization and hardware interoperability exists with the Windows operating system. There are some desktop computer operating system alternatives (e.g., Apple's Mac-OS and Linux), but Microsoft has successfully positioned Windows as the dominant PC operating system.

From the standpoint of an individual consumer, choice and competition are the two primary benefits of the interoperable Windows operating system:

- *Choice:* There are many PC alternatives that vary in terms of price, performance, sales support, and technical support.

- *Competition:* Choice breeds competition, which pushes PC manufacturers to innovate and push the limits on hardware price and performance.

While the Windows operating system is open to any and all hardware providers, it is still proprietary. And Microsoft has fought vigorously (some would say unfairly) to ensure that Windows dominates the intermediate and application layers of the desktop PC. As was the case with Big Blue in the 1980s, competition in the intermediate layers has been limited.

The Internet and Web: Open Standards

The Internet and the World Wide Web were founded on a series of nonproprietary intermediate-layer standards developed entirely through the efforts of an open, collaborative community. Anyone on the planet with appropriate expertise and interest could participate in Internet and Web standards development. An individual's employer had no bearing on his or her ability to participate, and no single organization owned the resulting standards.

Web and Internet standards can be freely used by anyone, in any new technology – no strings attached. Voluntary adherence to these standards permits any prospective Internet technology provider to create Internet and Web technologies. This is fundamentally different from the proprietary platforms of the IBM mainframe and the Microsoft DOS and Windows operating systems – where key technologies have been tightly controlled by corporate giants.

The surge in innovation and the adoption and deployment of new technology during the 1990s resulted largely from the Internet and Web open standards movement. The magnitude of this phenomenon may not be replicable in phase 2. We'll explore some of the reasons why in coming chapters. First, you need a basic understanding of how and why the Internet and Web open standards process emerged.

Internet Standards and the Public Internet

The public Internet is the most flexible platform for human communication in the history of mankind. It was made possible by people who wanted nothing more than to create a great computer network. A collaborative community of scientists and hackers created a set of standardized data formats and network communication and transmission rules and procedures, called *network protocols.* The participants used these protocols to interconnect a mix of heterogeneous networks. The resulting network of networks was known as an *internetwork* or internet.

Eventually, a single, large network of networks, called the Internet, emerged as a useful communications platform for researchers and academics.

As internetworking protocols evolved, a standards body – the Internet Engineering Task Force (IETF)* – was created to coordinate new internet standards. The IETF remains a collaborative, committee-based, not-for-profit organization that continues to churn out internetworking standards.

The Transmission Control Protocol (TCP) and the Internet Protocol (IP) are the core network protocols on which most Internet transmissions are based. The Internet Protocol handles Internet addressing, the TCP protocol ensures reliable data transport. Additional protocols standard-

* For more information see http://www.ietf.org.

ized by the IETF enable Internet services and coordination of the Internet infrastructure (e.g., routing protocols). Here are a few examples:

Protocol	Role
Hyper-Text Transmission Protocol (HTTP)	Formatting and transmission of Web content
Simple Mail Transport Protocol (SMTP)	Email message routing
Real-Time Transport Protocol (RTP)	Synchronizes real-time media transmissions

With the advent of the Web, Internet use began to take off in the mid-1990s, and it quickly morphed into an open, public, and global internetwork accessible to corporations, government agencies, not-for-profits, K-12 schools, and consumers. As public Internet use exploded, IETF protocols evolved with ever greater corporate interest and participation. TCP/IP and other IETF protocols were also integrated into a wider array of network products, and TCP/IP now interconnects an incredibly diverse mix of physical networks. Among the most common are:

- Ethernet networks over 10Base-T copper cabling, common in most corporations and other large organizations
- Metropolitan modem-, cable modem–, and DSL-based Internet access networks managed by local, regional, and international Internet Service Providers (ISPs)
- Regional, national, and global networks based primarily on high-speed copper lines, microwave links, and long-haul fiber-optic connections

Today's public Internet is a huge conglomeration of interconnected networks serving hundreds of millions of individuals and many millions of computing devices. Corporations have also embraced internet technologies, often securing their portion of the Internet behind a firewall to keep outsiders from gaining access to network, computing, and information resources. This sort of private corporate internet is called an *intranet*. An *extranet* is a private and secure Internet connection between business partners.

The open standards approach of the Internet has created something that a single corporate technology provider could only dream of. Even Microsoft, with all its clout, could never achieve near 100% penetration of Windows. But TCP/IP is implemented on desktop machines running the Windows, Linux, or Mac-OS operating systems as well as on servers running Sun Microsystems' Solaris, Hewlett-Packard's HP-UX, IBM's AIX, Intel-based PCs running Linux, and pretty much any other desktop or server computer running any type of general-purpose operating system. Even the IBM mainframe has been fitted with software for TCP/IP communication.

The Web

The World Wide Web has also been developed with layered technologies and open standards.

The IETF has some standards that dictate network transmission of Web content and interactions, such as the Hyper-Text Transport Protocol (see the table on page 84). However, most Web standards are not developed by the IETF, but by another standards body: the World Wide Web Consortium (W3C).* For example, the Hyper-Text Markup Language (HTML), the text-based markup language for defining hyperlinked Web pages,** is a W3C standard.

* For more information see http://www.w3.org.

** I offered a more substantial description of HTML, along with some sample HTML code, in the first book.

The W3C was created during the 1990s, and it is similar in operation to the IETF. W3C standards focus on Web data formats, languages, and interactions. It sets standards for new versions of HTML along with other Web-based languages and technologies (such as the Web services protocols introduced in chapter 11).

As is the case with IETF standards, W3C standards are totally open and free. Any technology developer has full access to all W3C standard specifications to guide design and programming of any new Web technology, service, or resource. This openness is fundamental to interoperability, which in turn ensures choice and innovation.

- While Microsoft Internet Explorer has indeed come to dominate the browser market, there are interoperable alternatives, such as the Opera Web browser and Mozilla.
- Multiple competing technology providers – including Apple, Microsoft, and Netscape* – all developed Web servers. But the Apache Web server is the number one Web server currently in use, and it was developed by yet another community-based collaborative effort, called the *open source movement* (introduced later chapter 9).
- The open standard of HTML has enabled significant competition, providing creator-collaborators with many powerful HTML editors – such as Netscape Composer, Macromedia Dreamweaver, and Microsoft FrontPage – to choose from.

The importance of the open-standards–based data formats, languages, and communication protocols of the Internet, the Web, and a host of other digital convergence technologies can't be understated. The first phase of convergence was a period where open, nonproprietary intermediate layer standards stimulated the rapid evolution and adoption of

* The Netscape server business is now owned by Sun Microsystems, which sells its Web servers under the brand name iPlanet.

new technology. From a technology industry standpoint, it seemed there were plenty of winners, and very few losers. Unfortunately, things aren't quite so rosy as we move into phase 2. Some gnarly issues are counter-acting the culture of the open-standards–based Internet. We'll explore several of these issues and barriers in the remainder of this section.

8. THE BUSINESS OF LAYERS AND STANDARDS

As we learned in the last chapter, technology layers and standards can have a significant impact on the evolution and adoption of new digital convergence technologies. For entrepreneurs interested in pursuing new technology and technology-based services and resources, this can translate into significant business opportunities. For incumbent providers, standards-based innovations can also disrupt revenue streams. Thus, as digital convergence has emerged as a major force in our society and economy over the past few years, standard technology layers have become a business battleground of sorts where winners and losers are made. Unfortunately, technology provider tactics now seem to regularly result in dysfunctional disputes, competing standards that fragment markets, and proprietary extensions that frustrate users. Let's look at a few examples to get a sense of how this plays out.

Handheld Devices

Personal Digital Assistants (PDAs) are small, handheld computing devices that you can use to manage and manipulate information and access the Internet. PDAs have actually been around for some time. The first time I saw one was in the mid-1990s – it was an Apple Newton, which was ahead of its time and is no longer in production. Over the past couple of years PDAs from Palm Computing, Hewlett-Packard, Handspring, and others have become popular. They are useful for salespeople and business professionals on the go, and there are several applications popular within specific industries (for example, in the medical services industry PDAs are used to record and access patient data).

Another type of handheld computing device that is also growing in popularity is the smartphone. This is a cellular phone with on-board general-purpose computing capability. Smartphones lack the raw computing power of a PDA, but they offer a viable alternative for lightweight handheld applications such as messaging, gaming, and Personal Information Management (PIM, which includes applications such as calendaring and address book access and management). PDAs and smartphones are now approaching a critical mass from the point of view of software developers considering an investment of time and resources to create new handheld applications.

Applications software developed for a particular computing device is usually written in a language such as C or C++ (Figure 1). Developing in C or C++ for one device and then porting to another is a time-consuming process, owing to the language features specific to the target operating system and device. If your target platform is the personal computer, choosing to develop using C or C++ for Windows has been a reasonable strategy. Without porting to Macintosh or Linux systems, you can reach a very large market of Windows desktops and notebooks.

Figure 1. Handheld Devices and Layers

Application Software	Application Software	Application Software
C/C++ Windows CE	C/C++ Palm OS	C/C++ Symbian OS
Ipaq. Jornada	Palm. Handspring	Nokia. Motorola

For application developers considering handheld applications, the choice of platform and language is not so straightforward. The handheld marketplace is rapidly evolving, and it is not yet dominated by Microsoft. The Palm OS is currently the operating system leader when it comes to the number of PDAs now in use, and the Symbian operating system is most common on cell phones. However, Microsoft is a significant potential player, and you can bet that it will pull out all the stops to move in and establish Windows as the standard OS – not only for Pocket PC devices, which typically run a version of the Windows CE operating system*), but for all other types of handheld devices as well.

With many different types of handheld devices on the market, should software developers program in C or C++ for a single platform? Or spend the time (and money) to develop C or C++ for one platform, and then port programs to other platforms? This can be a difficult and risky decision. Fortunately, there is another alternative: Java (Figure 2).

Figure 2. Java for Handhelds

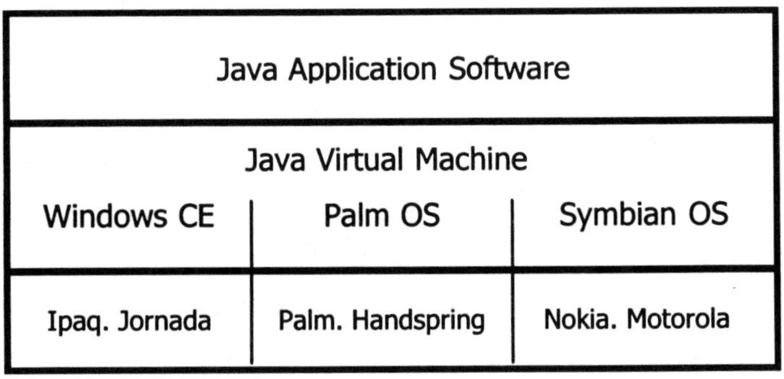

Java is a network-aware programming language developed through an open community process championed by Sun Microsystems.** A Java

* For more of Windows operating systems for handheld devices, see http://msdn.microsoft.com/embedded/choose/default.aspx.

** Initial Java development efforts were focused on consumer electronics applications. In the mid-1990s the original team tried to sell cable television companies on their processor-independent software development language, but these companies didn't want to give up control over their network and the consumer electronics devices connected to the network. The team turned its focus to developing for the Internet, demonstrating a Java-based browser called HotJava, which eventually led to widespread use of Java on the Internet. From Jon Byous, "Java Technology: The Early Years," *The Source for Java Developers* (http://java.sun.com/features/1998/05/birthday.html). Java has come full circle as the new version of Java is once again focused partly on network connected consumer electronics devices.

Virtual Machine, which can interpret and execute Java programs, has been implemented in a consistent manner across the operating systems of a variety of handheld devices. Thus, Java has become a viable platform-independent layer for developing handheld applications.

The promise of Java is optimistically characterized as "Write Once, Run Anywhere." In practice, most Java programs require some minor modifications to run on multiple platforms. Still, when compared to C or C++ programming, Java developers can write a single program, and then tweak it to run on multiple platforms with relative ease.

While Java is not standardized by a formal standards body, the Java community process and the focus on platform independence and interoperability combine to make Java a "standard" layer for the application developer community.

The most recent version of Java is available in three editions:

1. Java Standard Edition (J2SE) for desktop computers
2. Java Enterprise Edition (J2EE) for enterprise servers
3. Java Micro Edition (J2ME) for embedded and mobile devices

Java-based handheld software development has been a fairly popular approach over the past few years, and in 2001 Evans Data Corporation reported that J2ME was the top choice for wireless application developers.*

This example provides an opportunity to highlight a few of the realities that drive adoption and industry support of standard technologies:

1. Standard technologies can ensure a critical mass of users.

Handheld devices are in the early-growth stages of technology evolution and exhibit a typical early-growth characteristic: a number of alternative platforms with no clearcut market leader. By choosing Java to develop handheld applications for multiple operating systems, developers

* Lenny Liebmann, "Java on the Go," *InternetWeek*, May 23, 2001 (http://www.internetweek.com/indepth01/indepth052301.htm).

can reach a larger market and be protected – at least in the short term – from the prospect of the eventual market decline of any one platform.

2. Standard technologies aren't always superior.

While standard technologies offer some significant advantages, they sometimes raise some important technical considerations as well. In the case of Java, performance can be an issue. With C or C++ programming, a completed program is converted to an executable object in the native machine code of the platform it was developed on – so it is quick to execute. Java programs are distributed to target computers in symbolic "bytecode" objects, which require additional translation on each target platform. When a user needs to run a Java program, a special program called a Java Virtual Machine (JVM) must first convert the bytecode object to native machine code, introducing a performance penalty in the process. Also, Java programs aren't always stored locally – browser-based Java programs are downloaded prior to execution, draining available bandwidth and further delaying execution. These performance considerations are more or less important depending on the nature and complexity of a particular application.*

3. Technology standards can level the playing field for competitors.

Large corporate technology providers are often conflicted when it comes to standards. While vendor support for a given standard can level the playing field, stimulating a technology market in the process, it can also eliminate a proprietary competitive advantage. Technology vendors clearly recognize the overall value of open standards, and they will almost always profess to support open standards – although they typically

* For more on Java technical issues for handheld devices, including a discussion of nonstandard profiles for various handheld device types, see Carl Zetie, "What Ever Happened to Baby Java," *InformationWeek*, December 24, 2001 (http://www.informationweek.com/story/showArticle.jhtml?articleID=6508185).

will refuse to fully support any standard that may contribute to an eroded market share. For example, in the case of handheld devices, Microsoft shipped its Windows CE 3.0–based Pocket PCs without a Java Virtual Machine (preferring instead that developers use Microsoft application development tools). Windows CE–based PDA vendors had to add a Java Virtual Machine from a third party onto their new Pocket PCs – or require customers to install the software – so that end users could use increasingly popular J2ME applications.*

Let's review a couple of other standards issues taken from my own personal experience to help illustrate some of the other realities that can impact the ability of open standards to thrust new technology forward.

Streaming Media

Streaming media technology enables on-demand and live transmission of audio and video across the 'net. It is one of those technologies that is bringing us closer to the time when rich media will become integral to our Internet interactions. I've been creating streaming media content for some time now, and I've also had some great opportunities to do hands-on evaluation of competing streaming media technologies.

When you access a streaming media resource on the Internet, several seconds of incoming audio or video is downloaded onto your hard drive in a temporary file, called a "buffer," before your media player software

* BSquare claimed to be the first to offer a Java Virtual Machine for the Windows CE–based Pocket PC 2002 platform. From a BSquare Press Release, October 4, 2001 (http://www.bsquare.com/about/press/story.asp?PressID=275).

Hewlett-Packard bundled its own ChaiVM Java Virtual Machine on its Jornada Pocket PC. From Rich Hall, "HP Jornada 720: Still King of the 'Handheld' Hill," *PocketPC Magazine*, January 2001 (http://www.pocketpcmag.com/Jan01/jornada720.stm).

begins to play the media file (from the buffer). This buffering scheme can mask Internet congestion and transmission delays, so long as disruption isn't so severe that the buffer is depleted during playback (see Figure 3).

Figure 3

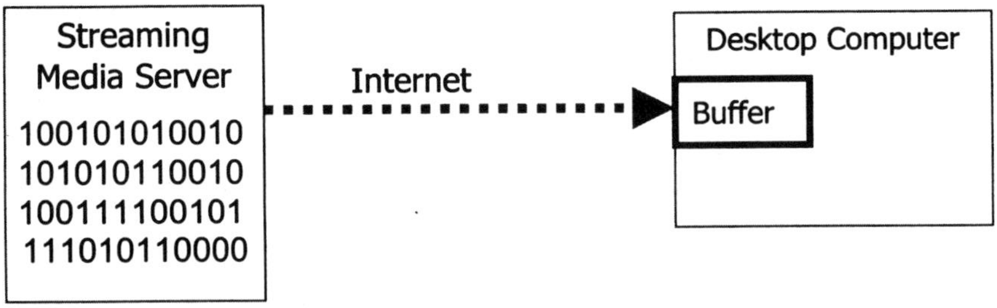

Standardization vs. Competition and Proprietary Innovation

> "We feel it's absolutely necessary for the industry to standardize on MPEG-4 and leave the holy war among RealPlayer, Windows Media Player, and QuickTime behind."
>
> Jim Wood, Vice President of AT&T Broadband*

Sometimes the standardization process is just too slow.

Uncompressed audio and video media representations can require an enormous volume of bits.** While video and audio compression standards MPEG-1 and MPEG-2, developed by the Moving Pictures Experts

* Junko Yoshida, "Cable Carries MPEG-4 Flag in Codec War," *Electronic Engineering Times*, May 12, 2000 (http://www.techweb.com/wire/story/TWB20000512S0015).

** I referred to voluminous audio and video binary representations as the "data volume problem" in the first book. In the first book I also described some of the compression techniques commonly used to decrease data volume for audio and video.

Group (a working group of the International Standards Organization), have been in use for many years, they were not developed specifically for Internet streaming media, and neither could effectively support low-bandwidth streaming. During the Internet explosion of the 1990s no viable compression standard was on the horizon, so the big three streaming media technology providers – Microsoft, Real Networks, and Apple – created proprietary compression technologies for their respective streaming media platforms. Real Networks and Microsoft developed their technology in-house, while Apple incorporated compression technology developed by another firm, Sorenson Media.

Since then, the rapid-fire evolution of this technology has had a dramatic impact on the quality of streaming media. We started using streaming video in the Whitman School back in the mid-1990s, and there is a huge difference between our early postage-stamp video and today's full-screen streaming video content.

During the late 1990s and into the 2000s, compression technology emerged as a key competitive feature of the streaming solutions from the big three. Each touted its compression technology as superior in some way.

I did a side-by-side comparison of Microsoft Windows Media and Real Networks streaming media products for *Network Computing* magazine in 2001 (I also looked at Apple QuickTime, but this was a PC-based review and QuickTime didn't stack up well – which resulted in my first hate email, from Macintosh zealots). I would have to say that Real Networks' compression was clearly superior when I used their innovative "two-pass encoding." Of course this was a simple, unscientific test – but the fact remains that proprietary compression innovation has been an ongoing area of fierce competition in the streaming media marketplace.

Of course there is a downside to this: incompatible, un-interoperable solutions.

- As a streaming media end user, you have to have the right player to view each different type of stream. While Microsoft bundles the Windows Media Player with Windows, PC users typically have to download and install the Apple QuickTime player or the Real Player. Macintosh users get a QuickTime player, but may have to download and install the Real Player or the Windows Media Player.

- For a streaming media developer or content provider, choosing to support a single platform may limit the audience. Supporting multiple platforms will cost you in terms of time and effort, but will allow you to reach a broader audience. This is not an easy choice to make.

MPEG-4 compression was standardized in 2000.* MPEG-4 incorporates many of the innovations introduced by the big three over the past several years. It also introduces some new features, such as the ability to interact with various media elements (e.g., you could turn off the video and just listen to the audio, or toggle a closed captioning text stream on and off). Microsoft, Apple, and Real Networks were actively involved in the development of MPEG-4. However, the market control end-game saw the big three slow to integrate MPEG-4 technology into their products and unwilling to abandon proprietary innovations.

Yes, additional home-grown innovations and improvements can improve the user experience; but the resulting lack of interoperability hurts the industry by complicating creation and delivery of streaming content, and constraining widespread end-user adoption of this exciting and very useful communications platform.

Proprietary Extensions and Competing Standards

The streaming media industry was involved in another intermediate-layer standards battleground in the late 1990s: technology that enables

* For more on MPEG-4 see the MPEG-4 Industry Forum Web site at http://www.m4if.org.

the synchronization of streaming audio and video with other media – including images, animations, text, and Web interactions.

Proprietary technologies for doing this in a relatively crude fashion have been available for some time. Microsoft included an encoder option for pushing Web pages to end users during live or on-demand streaming audio or video. RMEvents was the Real Networks solution, and it too enabled Web page push throughout playback of an on-demand audio or video stream.*

In the late 1990s Microsoft and Real Networks both pursued new mechanisms for synchronizing audio and video with other media, providing classic examples of *proprietary extensions* and *competing standards* in the process.

- Real Networks led the development of a W3C standard called Synchronized Multimedia Integration Language (SMIL, pronounced “smile”). Apple QuickTime and Real Networks both supported the SMIL 1.0 standard, but Real Networks introduced several non-standard features in its implementation. Thus, the QuickTime player was unable to reliably render a Real Networks SMIL stream. This is a good example of the fairly common practice of adding proprietary extensions to secure a competitive advantage and capture and retain market share. Unfortunately, this is yet another stumbling block on the road to standards-based streaming media interoperability.

* These technologies are often used with Web frames, which are sections of a Web window. One frame is typically used for an embedded streaming media player, while another is used to push Web pages with graphic images. The user sees an integrated interface that includes a media player together with continually updated Web pages that are synchronized with the audio or video stream. Often the Web pages are embedded PowerPoint slide images, creating an audio- or video-narrated slide show effect.

- When Real Networks was pursuing the SMIL 1.0 standard, Microsoft declined to participate. Instead, it was pushing an alternative standard which was initially called HTML+Time. Microsoft argued that synchronized media and interaction is not specific to streaming media – it is an overarching issue that should be addressed within the browser. With HTML+Time, Microsoft furthered its efforts to place Internet Explorer at the center of Internet interactions.*

Here again, the open standards ideal was at odds with the short-term strategies of large intermediate-layer technology providers, and not surprisingly, Microsoft was involved. Microsoft is the king of the hill when it comes to the intermediate-layer standards wars. But let's be fair about this: technology providers in all technology sectors and at all levels play the game. There is big money to be made. Microsoft is just very, very good at it.

The success of the open-standards–based Internet and Web demonstrated the broad and deep impact of open, cooperative technology development. But the open standards process is no longer an idealistic, collaborative, community-based effort. Now the process of creating, supporting, and extending intermediate-layer standards is increasingly tainted by the profit-oriented strategies and tactics of bottom-line–oriented corporate technology providers.

* All of the major streaming media companies, including Microsoft, have since joined forces in the development of the SMIL 2.0 standard.

9. Microsoft Tactics and the Open Source Movement

When you buy a personal computer, what do you look for?

A fast processor and lots of memory?

Advanced multimedia features?

A good warranty and responsive service?

Low Price?

These are typical of the things an individual PC consumer looks for. Manufacturer brand name may be a consideration, but by and large it is just one of many.

Now, what do you look for in an operating system? Or in an office productivity suite that includes a word processor, spreadsheet, database, and presentation software? Here, if you're like most of us, you're looking for Microsoft software, plain and simple.

In the intermediate and application layers, Microsoft has been very effective in gaining and sustaining widespread adoption of its PC operating system software, application software, data formats, and technical standards. Yes, I know some of you do look for Apple, Linux, or Corel instead of Microsoft, but if you look at the desktop computer marketplace as a whole, Microsoft has near-total control of the most often used aspects of these layers – with its Windows operating systems, Internet Explorer browser, and MS Office suite.

Microsoft has also launched a successful assault on the server marketplace over the past several years, challenging Novell in the realm of LAN-based file and printer sharing, and Sun, HP, IBM, and others in the application server market (application servers provide Web, email, streaming media services, etc.).

Microsoft.NET is Microsoft's new technology strategy and architecture designed to perpetuate Microsoft domination in the intermediate and application layers in the second phase of convergence, and beyond. .NET can enable seamless distributed computing and information sharing across a heterogeneous mix of network-connected desktops, servers, and mobile and embedded computing devices (e.g., smartphones and television set-top boxes). We'll explore .NET in more detail in chapter 11. For now I just want you to recognize that owning the intermediate and application layers has been and continues to be a core part of Microsoft's strategy.

Figure 1. Technology Company Revenues and Profits*

	2001 Revenue (in billions)	2001 Profits (in millions)
Sony**	$66.2	$152
HP & Compaq***	$78.8	$(–377)
AOL Time Warner	$38.2	$(–4,921)
Microsoft	**$25.3**	**$7,346**

As you can see from the above table, Microsoft's business strategy has proven very lucrative. Generating huge profits from the sale of intermediate and application layer software solutions – which are inexpensive to reproduce and distribute in large volumes – is obviously a very good business model. Let's take a look at some of the tactics Microsoft

* U.S. company data is from the Fortune 500 list on the Fortune.com Web site. Data is for the fiscal year ended on or before Jan. 31, 2002. All figures are for the year ended December 31, 2001.

** Sony data is from the Global 500 list on the Fortune.com Web site. Data is for the fiscal year ended on or before March 31, 2001. Profit figure is converted to U.S. dollars at the average exchange rate at Sony fiscal year end (Dec. 31, 2000).

*** Pre-merger figure, calculated by summing HP and Compaq figures.

has used to help gain and retain a firm grip on intermediate- and application-layer markets to deliver these kinds of numbers.

Microsoft Tactics

Strategic Use of Nonstandard Technology

In the early days of the Web, Netscape Navigator was the most widely used browser. Early on, Netscape developed a plug-in architecture that enabled third-party software developers to implement software capability right within a browser page (using <embed> tags), and this technology quickly became a de facto standard.

Microsoft introduced Internet Explorer once Gates and company woke up to the Internet and Web phenomenon. Microsoft implemented the plug-in capability within Internet Explorer, but it also developed an alternative, called ActiveX Controls, which incorporates proprietary technology developed for the Windows operating system. Initially, ActiveX worked only in the Windows Internet Explorer browser. Unix and Macintosh users lacked this proprietary extension.

Over time, Microsoft used its considerable resources to improve IE, introducing other proprietary extensions in the process. For example, in a subsequent version of IE, Microsoft added proprietary functions to its implementation of Dynamic HTML (an eventual W3C standard suite of technologies that offered significant new capabilities).

While Microsoft was pursuing its proprietary technology agenda, it was also using a variety of other tactics, such as bundling and PC vendor arm twisting (described in the next section), and eventually it won the browser war.

Since then Microsoft has used its hard-won clout with the IE browser to move the Web in the direction of Microsoft technology. For example, new versions of IE do not support the once popular Netscape plug-in architecture.

Microsoft has also set its sights on Java. Proprietary object-oriented (Java-like) programming languages, C# and J#, have been introduced as part of Microsoft's Microsoft.NET technology – languages that the technology trade press has referred to as "Java killers."* Also, when the Windows XP operating system was first introduced it did not ship with a Java Virtual Machine, which is required for Java programs developed for the Web browser.**

Microsoft's goal has always been to tie the browser to Microsoft-specific functions and development tools – which is great for Microsoft-oriented software developers, but counterproductive for Web developers in the Netscape (now iPlanet), Java, and Unix camps. This has infuriated many programmers and software engineers.

Bundling

Bundling refers to the Microsoft practice of including free software with its operating system, a tactic that has proven very effective in promoting the use of Microsoft software.

During the 1990s Microsoft began to work with major hardware providers such as Dell, Compaq and Gateway to have Internet Explorer installed as the default browser on the desktops of new computers. But Microsoft went further than that. It allegedly pressured hardware providers so that IE was the *only* browser installed. This enabled what many believe was an unfair advantage over the competition, and it, along with some other tactics that appeared unfair to many, attracted the attention of

* Andrew Orlowski, "Microsoft Describes Its Java Killer," *The Register*, June 27, 2000 (http://www.theregister.co.uk/content/1/11606.html).

** Aaron Ricadela, "Microsoft Nixes Java in XP, Draws Fire From IT," *Information Week*, July 23, 2001 (http://www.informationweek.com/story/showArticle.jhtml?articleID=6505925).

the Department of Justice, which eventually filed antitrust complaints against Microsoft.*

Bundling on the desktop does not begin and end with IE. Microsoft also bundles a variety of additional Microsoft applications with the Windows operating system, including its Windows Media Player and Windows Messenger.

While much has been said and written about Microsoft's desktop bundling practices, its server bundling practices are significant as well. On the server, Microsoft bundles the Internet Information Server (IIS) Web server along with Windows Media Services, its streaming media server software. Let's take a quick look at some of the pluses and minuses of Microsoft's bundling tactics in the context of streaming media.

Bundling and Streaming Media

In the Whitman School of Management we have used streaming media for live Webcasts and on-demand content delivery since 1997. Over the years I've also written about streaming media for *Network Computing* and for the streamingmedia.com Web site. Throughout this time I have had lots of hands-on experience with Microsoft's Windows Media technology, and I witnessed the evolution of Microsoft's Windows Media Services bundling tactics.

While I've been generally unimpressed with Microsoft's streaming technology, Microsoft has seen a steady growth in market share, and it is now neck-and-neck with Real Networks (see Figure 2). I'm convinced – based on what I've experienced, seen, and heard within the industry – that bundling has been an important factor.

* The "Court's Finding of Fact" in the U.S. vs. Microsoft case can be found online at the Department of Justice Web site (http://www.usdoj.gov/atr/cases/f3800/msjudgex.htm).

Figure 2. Nielsen-Netratings Streaming Media Audience Measurement

Unique Audience (April 2002)*	At Home	At Work
RealMedia - audio/video	17,059,000	11,633,000
Windows Media - audio/video	15,148,000	12,223,000
Apple QuickTime - audio/video	7,278,000	5,021,000

Bundling has a bright side. Streaming developers find that if they use a Windows-based server, Microsoft's free Windows Media server is bundled. They just enable and configure it, and go about the business of encoding and streaming content. And with Media Player pre-installed on every Windows PC, Windows Media developers can introduce streaming media resources confident that a large audience of prospective users will be able to access the streaming content without needing to download a new player. This is very attractive, especially among creator-collaborators with limited budgets.

Of course, you can also download and use a free streaming server from Real Networks, called the Helix Universal Server. But it allows a limited number of concurrent streams, and it does not support live Webcasting. Adding concurrent streams, live Webcast support, and other features (such as propagating streams to multiple servers when serving large audiences) can run you some Real Dollars. As of March 2004 an entry-level Real Server (the "Helix Universal Server – Standard"), with additional capabilities and support for two to four megabits per second of concurrent streaming, cost $1,999.95.**

Advantage Microsoft.

* "Nielsen/Netratings Launches New Web Multimedia Format Report, Tracking RealMedia, Windows Media, and Quicktime," NetRatings Press Release, June 20, 2002 (http://www.nielsen-netratings.com/pr/pr_020620.pdf).

** From the Real Networks Web site (http://www.realnetworks.com/products/server/index.html).

While the Microsoft streaming solution is bundled (and free), there is also a significant downside. The biggest problem in my experience has been the inferior quality of the Microsoft solution. I did a detailed, hands-on, side-by-side comparison of the big three – QuickTime, Windows Media, and Real Networks – for *Network Computing* magazine in 2001.* In comparison to the Real Networks products, Microsoft Windows Media was clearly inferior. I encountered a steady stream of glitches and problems setting up and delivering Windows Media streams, and the documentation was poorly organized and often misleading.

- When using the Windows Media on Windows NT, I repeatedly encountered a message indicating that a new version of the software was available and asking if I wanted to download and install. However, the new media player wasn't available for Windows NT.
- To use the encoder on Windows NT, I had to install a new version of DirectX, a Windows operating system component. I unsuccessfully followed the directions to download and install the software, only to find out later from my technical contact at Microsoft that the download instructions on the Web were out of date. By then I had switched to Windows 2000.
- At one point I had created video and audio files that I was using for test purposes. I could open them and encode them with no problem with the Real Producer program. But I got an arcane error message every time I opened them with the Windows Media Encoder. I discussed this at length with my contact within Microsoft, and we never resolved it. Later I found a solution on one of the streaming media discussion forums (which was to download a low level .dll system program).

* Andy Covell, "Real Network Skins Apple, Microsoft," *Network Computing*, February 19, 2001 (http://www.networkcomputing.com/1204/1204f1b1.html).

Throughout this evaluation, I got support from a top Microsoft technical pro who focused exclusively on Windows Media – and I was still very frustrated. Most other Windows Media users receive much less attentive service.

In contrast to the Microsoft minefield, I had zero problems getting streams up and running using the Real Networks streaming solution.

Part of the problem with the Microsoft approach has been a propensity for Microsoft to enable software application services and features through shared Windows software components. With Microsoft applications, a complex web of software interdependencies is common, and this was the cause of many of the problems and glitches that I encountered. This differs from third-party software, which lacks this sort of complex, intimate application and operating system integration.

Microsoft appears to have recognized the growing disenchantment with these sorts of technical shortcomings. With Windows 2000 and Windows XP we have seen an increased emphasis on system reliability. I hope Microsoft further escalates the priority for technical quality and reliability as it develops and implements additional OS functions and application features on PCs and servers, and on a growing variety of mobile devices, Internet appliances, and embedded systems.

Of course for many die-hard tech-heads, it's too little too late. Microsoft's aggressive, annoying tactics and its all-too-often unreliable technologies have alienated cadres of programmers, systems administrators, media developers, and others involved in a wide range of software, content, platform, and systems categories. A community of grassroots geeks – who believe they can do a better job of developing software than any commercial interest, including Microsoft – has emerged. This community has used the technologies of convergence to take matters into its own hands, creating a new and potent competitor for Microsoft and other commercial software providers in the process. This is the open source movement.

The open source community offers a great example of what a community of creator-collaborators can accomplish with the right motivation and community-wide digital connectivity. While this may be the first such community to shake up a major industry, it surely will not be the last.

The Open Source Alternative

The original creators-collaborators of the convergence era are the hackers – programmers, geeks, and computer nerds – who shaped our digital environment. Together, these bright, energetic, and Internet-connected individuals have been a powerful force for many years. While the dark side of the hacker community focuses on invading remote systems and wreaking havoc on the operation of large organizations and the lives of individual computer users, the bright side has focused its efforts on free, unfettered access to software and digital resources.*

A couple of decades ago, the free software movement got a boost when Richard Stallman initiated an ambitious community-based software development effort where all new software was made available via the distribution concept of "copyleft."** Copyleft gives anyone permission to execute, copy, and modify a program, but no permission to add additional restrictions. Programmers, hackers, and geeks could create and modify software to fix bugs and extend functionality in collaborative programming projects – which resulted in useful software that was freely available to anyone for any purpose.

During the 1990s, the rapidly improving Internet facilitated collaborative, community-based development of free software. Software –

* For more on hackers, hacker mentality and culture, and the work that hackers do see Eric Steven Raymond, "How to Become a Hacker" (http://www.catb.org/~esr/faqs/hacker-howto.html).

** Richard Stallman, "The Gnu Project" (http://www.gnu.org/gnu/thegnuproject.html).

which is, in some sense, just another digital media type (like text image, audio, or video) – could be easily shared via the 'net. Web sites with information and answers to frequently asked questions (FAQs), Internet bulletin boards (Usenet News), and email discussion forums complemented software source sharing, enabling free software developers to come together as a broad, diverse, and highly functional virtual community. The result was a surge in the production of free software.

While free software was thought to be pretty darn good, many prospective business users were wary, for a variety of reasons:

- While the notion of "free software" means more than software without cost, it is true that free software is also usually free of charge – a fact that flies in the face of many decision-makers in the profit-oriented business community.
- In keeping with a familiar refrain, "you get what you pay for," corporate IT managers believed the quality of community-developed software was inferior
- Free software came without the normal commercial technical support. Support for the software was through interactions with the free software development community. Most business IT managers assumed this too was inferior.

In the late 1990s, around the time that Netscape decided to provide open access to its popular Navigator Web browser software, the term "open source" was hatched as an alternative to "free." This was an attempt to put a more business-friendly spin on community-developed software at a time when the free software movement was getting a boost, thanks to Netscape's decision. Since then, the open source movement has blossomed.

The open source community has demonstrated that it can develop and support high-quality software; and many IT pros, even in the business sector, now see open source software as superior to commercial software from companies like Microsoft.

High-quality and very successful open source programs are now available in many software categories. As of December 27, 2003, the largest open source software development Web site, sourceforge.net, was hosting a whopping 73,535 projects and had 760,224 registered users; another popular open source site, freshmeat.net, was home to 31,243 projects.

Here is a sampling of some of the most successful open source software:

- Linux – a general-purpose Unix operating system that is the most widely used OS on Internet servers
- sendmail – the most common email transport agent software on the Internet
- Apache Web server – the most widely used Web server in existence
- MySQL, Perl, and PHP – popular database, data access, and scripting languages used for creating dynamic backend Web programs
- BIND DNS server – the most popular name serving software on the Internet (it supports the Internet naming system, called the Domain Name System, or DNS)

Open source has become a significant competitor for Microsoft, especially on the server. Commercial Linux providers such as Red Hat and SUSE now offer shrink-wrapped Linux software with commercial tech support. Large technology providers now also support open software: IBM sells Linux-based servers and includes the Apache Web server in its Websphere product line; HP sells Linux-based servers; and even Sun Microsystems is now shipping some Sun servers with Linux rather than Solaris.

Open source proponents believe that open source software can compete on a wide variety of platforms. My TiVo Personal Video Recorder is a computer running a version of Linux, and Sharp Electronics has introduced a Zaurus PDA that runs the Linux operating system. On the PC, Linux operating systems are shipping with advanced open source graphical user interfaces (GUIs) such as K Desktop Environment (KDE) and Gnome, which rival the intuitive look and feel of the Windows and Macintosh GUIs. Linux users can also run a wide range of software (for example, Gimp, a sophisticated image editor that looks and acts like Photoshop; Freevo, a home theater and Digital Video Recorder program; and Open Office, an application suite that includes a spreadsheet, word processor, and a presentation program). Lindows is a Linux-based operating system that offers a GUI that looks and acts like Windows, and consumer heavyweight Wal-mart is now selling low-cost computers loaded with Lindows rather than Windows.

Earlier in this chapter I very briefly introduced Microsoft.NET, Microsoft's core strategy and technology for supporting emerging Internet-centric distributed computing and information access across a wide variety of devices. The open source community is extending its technology in a similar fashion, and the fusion of open source software and the latest version of Java provides a comprehensive suite of software that can match the scope of the Microsoft solution.* We'll explore this in chapter 11 when we profile emerging Web services architectures. Before that, we need to review the evolution of Web technology in the next chapter.

* Of course the innovative open source community is seeking multiple mechanisms to provide .NET alternatives. One noteworthy initiative is Mono, a project to create a free implementation of the .NET development framework that includes a C# compiler and a library of program modules called a *class library*. For more details see http://www.go-mono.org.

10. CLIENT/SERVER AND THE EVOLUTION OF THE WEB

The Martin J. Whitman School of Management is in the process of designing a new building. Once built, I believe it will be recognized as one of the truly great buildings on the Syracuse University campus. I'm fortunate to be on the school's building committee, and I can tell you that the architectural design process has been fascinating.

As representative of the eventual "users" of this new facility, the building committee has spent a great deal of time with the architects. We've tried to define the character and purpose of this new facility – as it relates to the mission and culture of the school and to the "student centered research university" vision of the university – and then translate that into a set of spaces, with adjacencies that will reinforce function while ensuring a smooth flow of traffic.

In the first book I described computer architecture as the way in which computer and network technologies are configured to provide functionality for end users, while facilitating information and technology sharing and management within and across organizations, communities, and groups of individuals. Like the architecture of the new School of Management building, effective computer architecture must also fit with the character, culture, and activities of the individuals served.

While the technology layering perspective is useful, architecture brings into view the geographic, organizational, and usage patterns that drive new models for organizing resources, services, and interactions. And as underlying computing, communications, and digital media technologies evolve and converge, there are some interesting architectural possibilities. In fact, as new technology and creator-collaborator activities and culture spread throughout our office workplaces, home offices, and living rooms, I find emerging computer architectures even more fascinating than the design process for the new Whitman School building!

In the first book I laid out computer architectures in a historical context, by describing the following architectural models:

- Mainframe computing
- PC computing
- LAN/WAN enterprise computing
- Client/server computing

When I wrote the first book it was not clear (at least to me) what architecture or architectures would dominate as convergence evolved. In phase 1, client/server architecture was the dominant architecture. But this architecture has its downside, and other architectures – such as peer-to-peer and Web services – were gaining momentum.

Now that some time has elapsed, I'm ready to offer some more substantive insight and commentary on emerging architectures for phase 2. In this chapter I'll provide some background on the phase 1 client/server architecture, and bring you up to date on the evolution of the Web. In the next chapter I'll profile and comment on some alternative architectures that will have a significant impact in phase 2.

Client/Server: The Dominant Phase 1 Architecture

Client/server computing is an architecture that sees software on multiple desktop computer "clients" interacting with a single server computer to obtain shared resources, services, and interactions. Web, email, streaming media, and other servers provide this sort of access to shared information, computing resources, and interactions for personal computer clients across the Internet infrastructure.

While the client/server architecture works well for many applications, it is far from perfect. Here is a quick summary of three of the most significant problems and constraints of the client/server architecture as implemented during the first phase of convergence:

1. Performance bottlenecks.

Many Internet client/server technologies are accessible to a global audience. Demand is often unpredictable, and a sudden increase in concurrent use can effectively disable a shared service or resource. Potential client/server choke points include networks and servers where network traffic and requests for service or processing are aggregated. Technologies such as load balancing servers have been developed to deal with these bottlenecks, but sudden traffic spikes still flood networks and overloaded servers. Some of the alternative phase 2 architectures profiled in the next chapter were developed to solve this problem.

2. Costly client and server administration.

Client software applications seem fairly simple to some, but for many, installing, configuring, and using client software is cumbersome and time-consuming. In many corporate settings systematic, tightly controlled mechanisms are used by IT pros to administer client software across the enterprise (described in chapter 4). Administering servers requires IT skill and expertise that is also beyond what some creator-collaborators possess, and outsourcing for this expertise can quickly drain limited financial resources.

3. Limited content sharing.

While the centralized server of the client/server architecture enables efficient, managed resource sharing, the ability to easily share resources among Internet users is sometimes constrained. This is often a function of centralized control, where server administrators constrain or disallow access to functions or information, citing resource or security concerns. In other cases the server owner flat-out refuses to provide a particular type of interaction or resource for any number of reasons. And sometimes demand for an interaction or resource exceeds the ability of a client/server infrastructure to cost-effectively service that demand. These factors stimulated the development of peer-to-peer content sharing programs such as Napster, Kazaa, and Morpheus (described in the next chapter and again in Part IV).

In the first book you may recall that I proclaimed that the client/server-based World Wide Web was the foundation digital convergence application technology. Since then, the Web has seen tremendous evolution, and today's Web provides functionality that goes far beyond the original application that enabled global access to static, hyperlinked HTML pages.

The evolution of Web technology has led to an important new architecture called *Web services,* which I will profile in the next chapter. Let's first take a look at how the Web evolved to become a programmable platform used to develop and deliver a wide variety of new Web-based application technologies, resources, services, and interactions.

The Web: From Application to Infrastructure

The best ideas are very often simple, and the original World Wide Web was no exception. You could use a basic text editor to create simple HTML pages (typically containing text, images, and links to other Web pages), and then upload them to a Web server, effectively publishing them to a global audience of Web surfers (Figure 1).

Figure 1

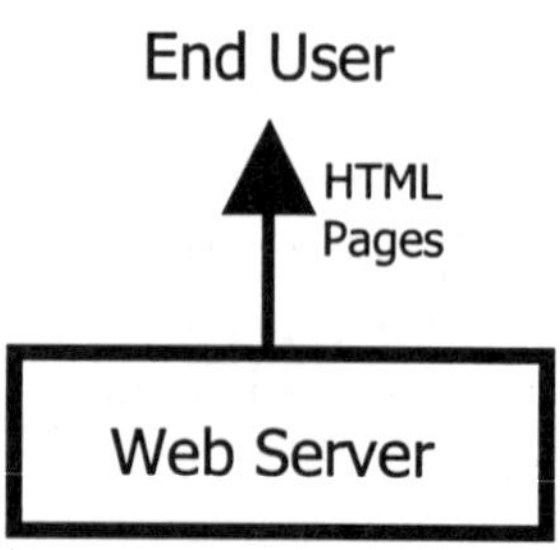

While this was useful, it wasn't long before Web functionality was extended. One of the first significant technological extensions to the Web was the Common Gateway Interface (CGI). This enabled Web site developers to create HTML forms that would take information from users and pass this information to "back-end" programs that could retrieve and update data files, and then dynamically compose HTML Web pages. One of the early and very popular implementations of CGI technology was a combination of the open source Apache Web server and an open source data access and manipulation language called Perl (Figure 2).

Figure 2

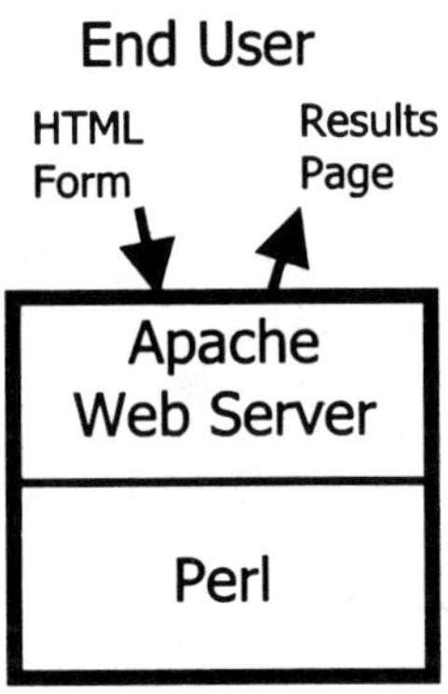

Web-based forms allow users to input information using checkboxes, drop-down menus, fillin blanks, and radio buttons. The input values are passed on as parameters to Perl programs, which dynamically generate HTML pages that are then returned to the Web server program, and then transmitted back to the user's browser. This sort of CGI capability is used to create Web-based interactions – such as e-commerce, online discussion, and distance learning – that have become quite common. Apache and Perl are still widely used. In fact, this is an option supported by Affinity.net, the Web hosting provider where I maintain digital-convergence.com.

Over time a variety of Web-server–based software application development environments – similar to Apache/Perl – emerged. At the School of Management we bought a solution that integrated our Netscape Web server (now called iPlanet after being acquired by Sun Microsystems) with a back-end programming language called ColdFusion. Unlike Perl, which manipulates information in flat files that reside on the Web server, ColdFusion has the ability to interact with an external database management server, such as a Microsoft SQL Server, which is what we use in the Whitman School (Figure 3). This is based on a network database access standard, called Open Database Connectivity (ODBC), which most mainstream database management systems support.

Figure 3

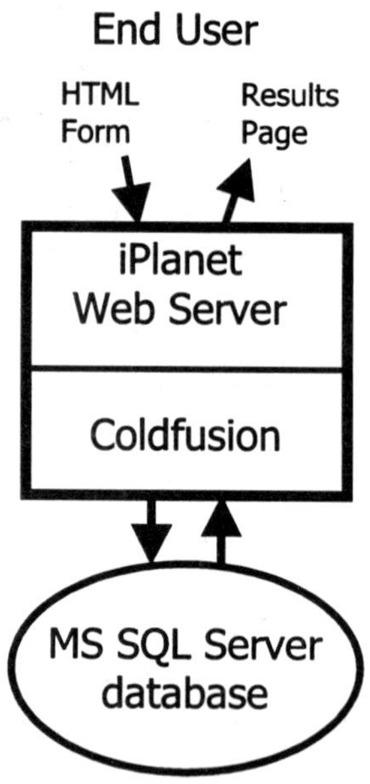

One advantage of this architecture is the separation of Web and database processing. This alleviates strain on Web server hardware since it isn't tasked with data access and management (as is the case with an

Apache/Perl server). The Web server does what it's good at, which is providing quick turnaround on requests for HTML pages; and the database server does what it is good at: retrieving, inserting, and deleting data items from a database.

This modularity also enhances server management: backups are quicker and easier to schedule; the database and Web server software are freed from some potential software conflicts; the servers can be tuned and upgraded independently; and a bottleneck in the database server (too many users generating lengthy searches or complex transactions) won't hamper end-user access to static Web pages.

Modularity in Web processing has been further extended in recent years to include Web-based application programs. Several vendors now offer this functionality on dedicated servers called application servers (see Figure 4). Some Web application servers support public e-commerce

Figure 4

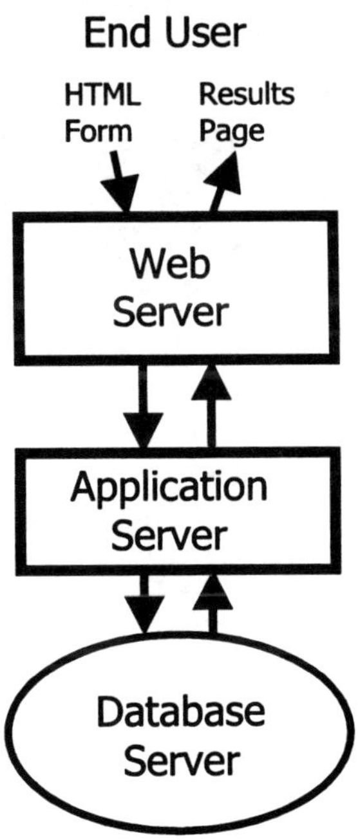

services and transactions, while others target in-house enterprise applications such as customer relationship management (CRM) and enterprise resource management (ERP) applications, which were introduced in chapter 4.

This architecture is often referred to as a three-tier architecture, because it consists of the Web server, application server, and database server tiers. Of course, like many things in the high-tech sector, the definition of a three-tier architecture is fuzzy. There are many variations to the three-tier theme, and when this was marketing jargon de jour, marketing types would fit almost any architecture to three tiers. And now vendors are touting *n*-tier Web-based architectures, providing even more flexibility in how these systems are presented and marketed to prospective customers.

Modularity has also enabled Web applications to scale to serve large numbers of users. A back-end computer called a "load splitter" can manage activity across a modular architecture of multiple Web, application, and/or database servers (Figure 5). This approach has become commonplace where high-traffic Web sites and Web applications are deployed.

As the Web has become increasingly modularized, a new category of software provider has emerged, called an Application Service Provider (ASP). An ASP creates consumer-oriented and business-to-business Web-based application software that is scalable and can serve large numbers of users across the Internet via a Web browser. Microsoft's Hotmail email application and Yahoo! Stores, which enables an entrepreneur to dynamically create and administer a Web-based storefront all through a Web browser interface, are just two examples of successful, creator-collaborator–friendly ASP applications.

Figure 5

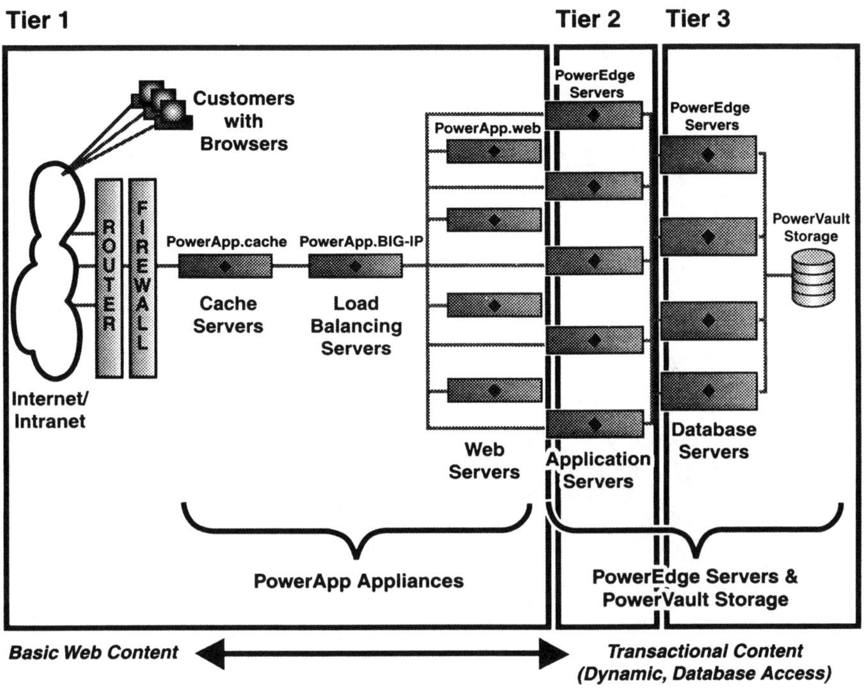

Some ASPs offer software to large organizations on a contract basis; others require an individual subscription; and some opt to offer the software for free and sell advertising to generate revenue. The ASP model for software delivery is a precursor to the new Web services architecture introduced in the next chapter.

This brief overview of Web technology evolution is meant to raise your awareness of the significant transformation of the Web – which has become an infrastructure for creating sophisticated and scalable client/server-based software, resources, services, and interactions. While you and I can still put up static HTML pages, application providers and IT professionals now view the Web as a fertile, sophisticated platform for creating Web-based technologies. Heavy hitters such as Oracle, IBM, Sun Microsystems, Microsoft, and HP all offer intermediate-layer Web technologies that constitute powerful platforms for developing Web-based applications.

In the next chapter we'll turn our attention to three emerging architectures that will play a significant role in the second phase of convergence.

11. Emerging Phase 2 Architectures

The client/server architecture, profiled in the last chapter, has been the dominant architecture of phase 1, and it will continue to provide us with shared resources, services, and interactions in phase 2. There are three other emerging architectures that will play a considerable role in phase 2. In this chapter I'll introduce each and characterize typical uses.

Web Services: The Enterprise Architecture

What kind of architecture will enable new forms of computing and network access that someday will become integral to the fabric of our everyday lives? For many who believe the Internet and the Web will remain central to the architecture, the answer is something called *Web services.*

During the first phase of convergence, the Application Service Provider model for Web-based software and service delivery (described in chapter 10) heralded the rise of a new type of Web-based computing architecture. In phase 2 an increasing range of mobile and rich media appliances – e.g., smartphones, PDAs, DVD players, and television set-top boxes – are being fitted with high-speed Internet access, becoming important components of a new distributed computing environment. The Web services architecture enables seamless, cooperative access to information and computing resources across this increasing range of interconnected devices.

Web services can take the form of client/server interactions or server-to-server cooperative processing; and they can be organized in multilevel tiers. For example, your PDA phone could request a Web service from your set-top box, which could in turn satisfy that request using information gained from a Web service on a remote Internet server.

The grand Web services vision is compelling, since it paints a picture of an interconnected world of functionality based on open standards. We know from preceding chapters that open standards can enable choice, competition, and innovation. However, we've also seen that the business of standards all too often constrains the creation and adoption of open standards (so be sure to have a grain of salt handy when reading the hype about open-standards–based Web services).

Web services does use a set of open-standards–based languages and directories to catalog, describe, and access Web-based applications and information (Figure 1).

Figure 1. Web Services Standards

WSDL	Web Services Description Language - Describe an Internet-accessible Web service
UDDI	Universal Description, Discovery, and Integration - Create registries of Web services
SOAP	Simple Object Access Protocol - Specify the formats and parameters used to access or execute a Web service

These W3C standards provide the "glue" that enables interconnected devices to publish, identify, and share information and Web-accessible applications and information resources. Here are a few more examples:

- Your smartphone could access a price quotation Web service on your corporate intranet, which would in turn call on a variety of complementary, integrated Web services to generate a quote.
- Your television set-top box may be able to access specialized video resources, purchase airline reservations, or execute a stock trade based on Web services made available by your cable provider, travel service, and financial services firm.
- If you take courses at a university, you may find a Web portal interface that enables you to customize access to a wide variety

of university Web services, from financial aid applications to library information retrieval services.

- From your PC Web browser, you could access a sports betting Web site which calls on a sports information Web service to provide up-to-the-minute information on games and players and a gambling transaction Web service for placing bets.

The Web services architecture is so compelling that Microsoft has redefined its core technology strategy around its Microsoft.NET branded Web services technologies. An alternative Web services strategy driven by the Java and open source communities is gaining considerable momentum. Here is a quick rundown on the activities and technologies of these two competing camps.

The Microsoft Solution

Microsoft has pursued an operating-system–centric strategy for many years, and in chapter 9 I presented some of the mechanisms, such as bundling, that Microsoft has used to gain and keep its lock on the PC operating system market and increase its presence in the server market. As a variety of new devices has emerged Microsoft has worked hard – although without the same degree of success – to make Windows a standard for PDAs, set-top boxes, and other mobile devices and Internet appliances.

With Microsoft.NET Microsoft signaled a significant shift in its strategy. Chief Software Architect Bill Gates recognized that in the age of convergence the network would be central to any successful computer architecture, not the operating system.

Microsoft has moved fairly aggressively with its .NET strategy. It has already launched a variety of services such as Passport, a user-authentication service that provides access to other .NET services (e.g., you now need a Microsoft Passport in order to use Microsoft's Hotmail service). Other early .NET services include document storage, calendaring, contact management, and an electronic wallet service. Microsoft is also pushing a variety of .NET development tools for third-party software and application developers.

Microsoft.NET Web services technology could ultimately enable Internet access to your software and your information, regardless of what device you're using and where on the 'net you're using it.*

Some new .NET services, such as Passport, have been poorly received by the media and many end users, including many creator-collaborators. People are wary of Microsoft's self-anointed role as a keeper of personal information. The company's track record, in terms of software reliability, privacy, and protection against security intrusions, has been far from stellar.

While the Microsoft solution uses open standards, such as SOAP, there are also a variety of Microsoft-specific technologies that are part of .NET. Visual Basic.NET is a proprietary scripting language, and Microsoft.NET services run on Windows-based servers. Microsoft's new Java-like programming languages (C# and J#) are targeted for Web services programmers. These new technologies are designed to ensure that Microsoft products are at the core of most Web services implementations – and many believe Web services will become the dominant phase 2 architecture.

* ".NET is a set of software technologies for connecting information, people, systems, and devices. This new generation of technology is based on Web services – small building-block applications that can connect to each other as well as to other, larger applications over the Internet." From the Microsoft Web site (http://www.microsoft.com/net/basics).

Of course, .NET has also become a very important Microsoft brand name, and as a result, according to Mark Miller, a colleague here at Syracuse University who has spent a great deal of time understanding Microsoft's .NET offerings, many technologies that have been around for some time and were not developed as part of a Web services strategy are also included under the .NET umbrella. As Mark puts it: 'everything good is called .NET.'"

Java-Based Web Services

Java is among the most popular programming languages of all time. It is portable and platform-independent. While not open source, the open, Sun-Microsystems–enabled collaborative effort to develop the Java language, called the Java Community Process, is somewhat similar to the community-based development process of open source software.

Java designers and developers collaborate in the development of new Java languages. When finalized, these languages are used to create something called a *reference implementation*. Technology providers then create Java products and applications that comply with the Java "standard" as modeled in the reference implementation. The end result is high-quality, platform-independent software – which ensures choice, competition, and innovation. Java also has a strong network orientation, in terms of features and functions, and it is popular among Web services developers.

Version 2 of Java includes the standard edition, for PC-based applications, and two other specialized language suites:

- *Java 2 Enterprise Edition (J2EE):* These are Java language components that are typically implemented as a bundle on a dedicated application server. J2EE components include Java Server Pages, Java Message Service, Java Database Connectivity, and Java Naming and Directory Interface.
- *Java 2 Micro Edition (J2ME):* J2ME is a set of specialized Java languages created for use on ultraportable devices, Internet appliances, and other embedded systems.

Web services protocols, such as SOAP, UDDI, and WDSL, can be used with Java 2 programs to provide Web services across a multitude of 'net-connected devices. Combine that with open source software, such as the Linux operating system and the Apache Web server, and you have a viable, comprehensive alternative to the Microsoft solution.

The reliability, power, and standards-based platform independence of the Java/open source combination has attracted major technology providers such as IBM, Sun Microsystems, Oracle, and HP. Many large technology-based service providers also welcome an alternative to the Microsoft solution, and they are collaborating on significant initiatives that are focused on Java, open source, and Web services.

The Liberty Alliance Project is one such initiative. The Alliance was created in reaction to Microsoft's introduction of its Passport service.* The Liberty Alliance is developing an alternative open standards solution for Internet identity management. Founding members include Java champion Sun Microsystems, America Online, Sony, MasterCard International, Real Networks, American Express, and HP.**

Web services is one of the next strategic IT industry battlegrounds, and the open source and Java communities – with the backing of several industry heavyweights – have combined to become a major force that Microsoft must now reckon with.

The Enterprise Embraces Web Services

The notion of open, interconnected Web services has been seized upon by enterprise IT professionals and the technology providers that serve enterprise IT customers. Many seem to view Web services as the next panacea for bringing enterprise applications and information sharing functions together within and across organizations. For example, enterprise portals will provide employees with integrated, customized access to business information and applications from throughout the enterprise; and B2B applications promise to leverage the seamless connectivity of Web services across secure Internet connections.

---

* George V. Hulme with Martin J. Garvey and Aaron Ricadela, "Liberty Alliance: No Passport Needed," *InformationWeek*, October 1, 2001 (http://www.informationweek.com/story/showArticle.jhtml?articleID=6506470).

** For more information see http://www.projectliberty.org.

I've characterized Web services as the *enterprise architecture*, since it has recently emerged as the new enterprise IT bandwagon. A May 2002 research report from IDC, an IT and telecom research firm, projected that Web services projects would generate $7.1 billion in the U.S. by 2006. A press release announcing the report quoted Sophie Janne Mayo, IDC's director of Wireless and eCommerce Implementation Services research: "Given the complexity of Web services, consultants and system integrators will be instrumental in forwarding the adoption of Web services among enterprises."* This is consistent with the corporate IT community's continuous focus on implementing very complex enterprise technologies, at considerable cost – and often with very mixed results (this is described in more detail in chapter 4). As a result you won't be surprised to learn that I remain a bit skeptical of enterprise Web services deployments.

The complexity and effort required to develop and implement Web services also places this architecture outside of the domain of the typical creator-collaborator, who will likely leverage Web services only as an end user.

Content Delivery Networks: The Mass Media/Mass Market Architecture

A Content Delivery Network (CDN) is an alternative computer architecture that emerged around the turn of the 21st century to support one-to-many dissemination of sophisticated Web pages and rich media content, such as streaming audio and video. With a CDN, content is distributed from a central source to local devices that stage the content on the edges of the Internet at local ISP points of presence, where it is made available to users. The local device is called a *cache* (or *edge server*),

* IDC Press Release, May 28, 2002, available at the WebServices.Org Web site (http://www.webservices.org/index.php/article/articleview/428).

and it is simply a computer that acts as a local content store. Web pages load quickly and streaming content is accessible without delay, since they're delivered directly from the local cache, obviating the need for transmission across the public Internet (Figure 2).

Figure 2

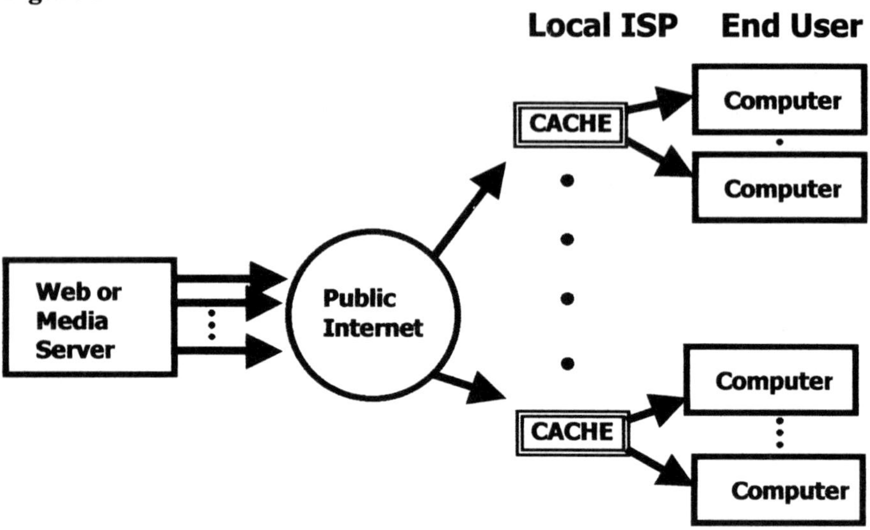

Companies operating heavily trafficked Web sites often contract with CDNs, such as Akamai or Speedera, so their Web pages load quickly in end users' browsers. Streaming media content providers also often contract with a CDN to ensure a good viewing or listening experience. Some progressive technology providers, like Real Networks with its distributed Helix server technology, offer caching distribution technologies for enterprise or network service provider customers who want to enable an improved end-user experience.

The Content Delivery Network was not conceived as a general-purpose architecture for distributed computing like the Web services architecture. But the CDN, which was once focused primarily on Web and streaming media content delivery, now provides support for edge-based software delivery as well. For example, Akamai now supports a markup language called Edge Side Includes for tailored page generation at the cache, and it has announced that it will soon support Java 2 and .NET processing at the edge.* This makes the CDN architecture useful for mass-market–oriented e-commerce as well.

While CDN-like caching technologies and techniques are used in the context of large enterprise intranets, the CDN is very appealing to companies who want to reach a large Internet audience. The CDN architecture can ensure a quality experience for a large number of concurrent users, in a tightly controlled fashion. Thus, I've lightheartedly dubbed the Content Delivery Network the *mass media/mass market architecture*.

Figure 3. A Sampling of Akamai Coporate Customers**

1-800-Flowers	L'Oreal
Apple Computer	Merck & Co.
Best Buy	The Motley Fool
BET Interactive	MSNBC
Charles Schwab	MTV Networks
The Chicago Sun-Times	NASDAQ
Columbia House	The New York Times
FedEx	Reebok International
Frederick's of Hollywood	Reuters
General Motors	The Sharper Image
Harley-Davidson	Toyota
Houghton Mifflin	Travelocity
J.C. Penney	U.S. Army Recruitment
John Wiley & Sons	Victoria's Secret
Lands' End	Warner Music

* Jennifer Mears, "MTV Gets 'Edgier' with Akamai Service," *NetworkWorld-Fusion*, August 19, 2002 (http://www.nwfusion.com/news/2002/0819carrier.html).

** From the Akamai Web site as of February 27, 2004 (http://www.akamai.com/en/html/about/customers.html).

Many creator-collaborators are focused on specialized, niche-oriented interactions, which don't require CDN technology. I can envision an alternative caching scheme that would be much more useful for creator-collaborators: Imagine a personal cache that sucks all sorts of content off the 'net and from broadcast sources, and is totally under your control (like a Personal Video Recorder – such as TiVo or ReplayTV – on steroids). You could slice, dice, integrate, and view this content any way you like. What would be the platform for this cache? I believe it may well be the desktop PC. And the best architecture for feeding this cache with diverse content from across the planet? Probably peer-to-peer.

Peer-to-Peer: The Creator-Collaborator Architecture

Peer-to-peer (a.k.a. P2P) is a loosely controlled, grassroots architecture that supports new forms of information sharing and collaboration.

The peer-to-peer architecture is very simple. It is based on the empowering notion that each computer on the network is both a client and a server. This enables a network of shared resources that can overcome client/server challenges including the potential for bottlenecks. P2P can scale to service many, many users without stressing any one network or server.

Napster, the breakthrough peer-to-peer application that was at the cusp of a wave of very popular peer-to-peer software programs, was developed by creator-collaborator Shawn Fanning. He was a college student at Northeastern University when he developed Napster to more easily and effectively share music files across the Internet. In combination with other peer-to-peer applications – such as instant messaging and the SETI@home initiative, which uses spare desktop computer processor cycles to assist in the search for extraterrestrial life* – P2P became a very

* For more information see the SETI@home Web site (http://setiathome.ssl.berkeley.edu).

popular architecture during the end of digital convergence phase 1. Another familiar P2P technology is instant messaging (covered in chapter 14), which is becoming multimedia-enabled in phase 2.

As we move into phase 2, P2P is being used in a wider variety of applications. For example, AllCast (Figure 4) and ChainCast are streaming media solutions based on the peer-to-peer architecture; and Groove Networks offers a P2P-based workgroup collaboration platform popular in many business settings.

Figure 4

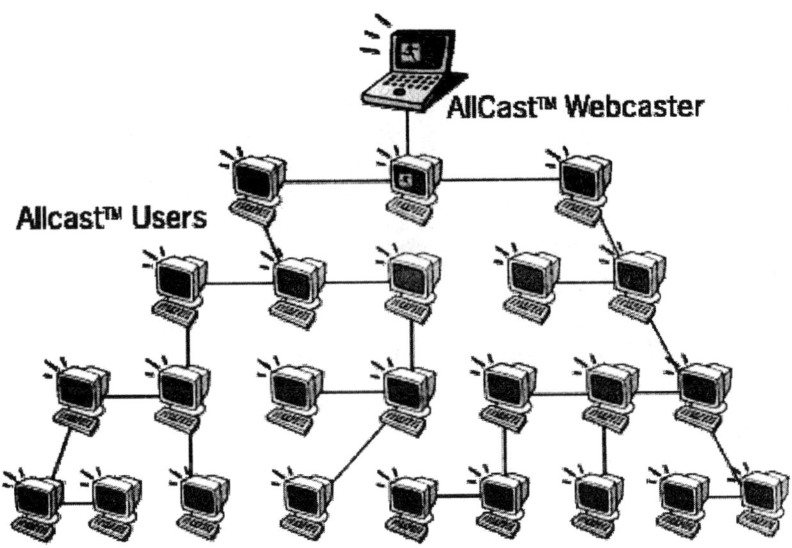

AllCast peer-to-peer streaming media
(© AllCast. Reprinted by permission.)

Music and movie sharing are still among the most popular P2P technology applications. At Syracuse University, we've seen bandwidth utilization for P2P music and video sharing skyrocket over the past several years (the university's network administrators responded by throttling P2P transmissions from SU computers to users outside the SU Internet connection).

The power of the peer-to-peer architecture is its ability to organize and leverage the vast computing and information resources that exist at the edges of the Internet – on my desktop PC and yours, and on the desktops of millions of other Internet users. And it does this without any dependence on centralized servers.* You can share resources, machine-to-machine and person-to-person, without having to inform any central authority (and without having to pay an access fee). What's more, that access is free from the bottlenecks of client/server and other centrally controlled architectures.

Many peer-to-peer applications are a part of the hidden 'net, or underground 'net, where totally decentralized computing means a lack of control over user behavior. While P2P enables us to share music, software, CPU cycles, images, and jokes, the dark side of humanity is also transmitted through peer-to-peer interactions (e.g., viruses, worms, and pornography). The peer-to-peer architecture is a good fit for creator-collaborators who seek to benefit from the unrestricted sharing of digital ones and zeroes and can deal with these rather unavoidable rough edges.

While many peer-to-peer technologies, such as Kazaa, are anathema to conventional content providers and some technology providers, P2P applications also sell hardware. Some large technology providers, such as Intel and Sun Microsystems, actively support this empowering architecture.**

---

* Some peer-to-peer applications, such as the original Napster, use servers to coordinate the connections between peers. Other P2P file sharing applications, such as Gnutella, avoid this server-based coordination to avoid the legal implications that result when copyrighted works are shared.

** For example, see Intel's online "Peer-to-peer Developer Center" (http://www.intel.com/cd/ids/developer/asmo-na/eng/technologies/peertopeer/index.htm?iid=techtrends+tech_peer_to_peer&).

P2P holds great promise for open access to a world of information and content, even as the second phase of convergence sees some large technology and content providers working feverishly to tighten their grip on consumer and creator-collaborator interactions and activities. We will explore these themes in more detail later, in section IV, when we profile a variety of compelling application technologies.

The next two chapters will provide an overview of the evolution of communications technology and profile the rapidly converging telecommunications and computer networking industries, which appear to be increasingly under the control of a small number of very large corporations.

12. INTERNET TECHNOLOGY CHALLENGES

> "The next decade . . . will see the net spread even further and start to become the basic communications infrastructure for almost anything."
>
> Vinton Cerf, January 1, 2004*

The Internet has evolved into a major communications platform that enables new forms of human and organizational interaction. While it should be no surprise that many believe it can indeed become THE network of Planet Earth, a series of difficult technical, industry, and market challenges stand in the way. In this chapter we focus on technical issues and barriers. We'll profile some industry and market issues and barriers in the next chapter.

Conventional Networks and the Internet

In the first book I pointed out that there are three different types of networks that most knowledge workers and consumers use on a daily basis. Here is a quick comparative recap.

1. Circuit-switched telephone networks

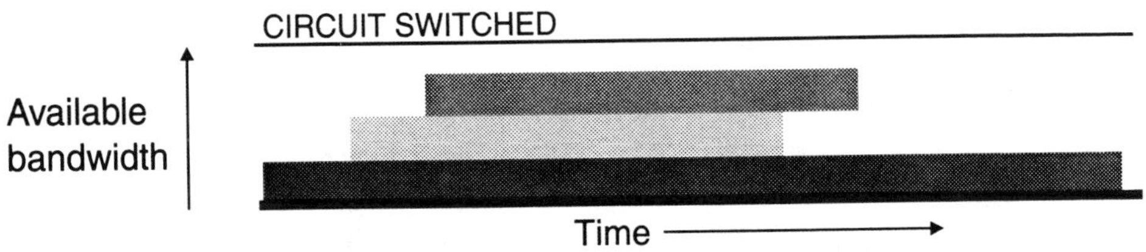

* From a BBC interview with the "Father of the Internet," Vinton Cerf. Mark Ward, "What the Net Did Next," *BBC News World Edition*, January 1, 2004, (http://news.bbc.co.uk/2/hi/technology/3292043.stm).

Your home phone has a wired analog electrical connection to a phone company facility called a central office (CO) where your phone wire connects to an automated switching device. When you place a call, this switch either routes the electrical signal to another local phone line or initiates a digital transmission across the interconnected digital networks owned by whatever mix of local and long distance carriers are involved in the completion of the call. When the digitally transmitted signal reaches the CO local to the party you're calling, it is converted back into an analog electrical signal and sent out across the phone line of the party you are calling.

The digital portion of today's telephone networks allocate a fixed amount of transmission capacity (i.e., bandwidth) for each and every call, from end to end across the network – ensuring high-quality connections and clear, reliable phone calls. The end-to-end reservation for each phone call is a virtual circuit, and the network is a called a *circuit-switched network*. The reliability ensured by the bandwidth reservation mechanism of this type of digital network is referred to as Quality of Service (QoS).

While we've all come to know and love the conventional phone network, it does have its drawbacks. Circuit-switched networks are inefficient, since considerable bandwidth is dedicated to each call, for the duration of each call, even when no one is talking. Also, conventional phone service does not provide for end-to-end integration of multiple media types owing to the local analog connection which is designed for straightforward conversion between electrical signals and sound waves at the phone handset.* The telephone network is also designed to support a limited number of endpoints in a single transmission (and most of us continue to use the network primarily for individual, point-to-point conversation).

* In order to use the phone network for digital transmission to the endpoints you must add modems to modulate and demodulate digital data across the analog electrical connection.

2. Broadcast networks

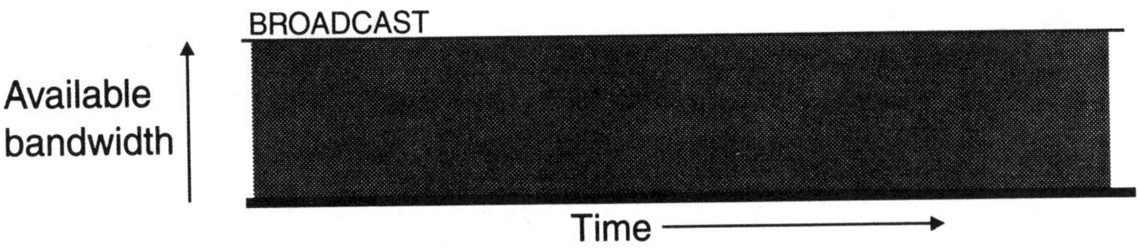

Local radio and television stations and cable and satellite television networks utilize all available bandwidth to transmit high-quality media in one direction to large numbers of recipients. Most of these broadcast networks have historically used analog transmission, although an increasing number now use digital transmission (e.g., satellite television and radio, digital cable, and HDTV broadcasts by local TV channels).

Like the circuit-switched telephone network, the conventional broadcast network has significant limitations. It has little or no capacity for interaction (thus, "talk-radio" must rely on the phone network and the Web for listener interactions). Media integration is also not an option: only audio and/or video are supported.* And while broadcast networks are very effective, there are some bandwidth utilization inefficiencies. For example, cable television networks concurrently broadcast all channels into our homes even though most of us usually watch only one program at any one time.

The broadcast network is also a tightly controlled scheduled content delivery mechanism. Listeners and viewers have no direct control over content transmission – the "what" and "when" of content delivery are decided by broadcast programmers. To receive desired content you must

* While this is not entirely true – for example, data can be transmitted in the vertical blanking interval of television broadcasts, a capability most often used for the transmission of closed captioning information – conventional broadcast networks were designed primarily for the one-way transmission of a single media type.

be available at the time that content is broadcast or you are out of luck (unless you use a local recording device such as a VCR or Personal Video Recorder).

As you continue to learn more about the capabilities and flexibility of convergence technologies, these limitations become more obvious and more frustrating.

3. Packet-switched computer networks, including the Internet

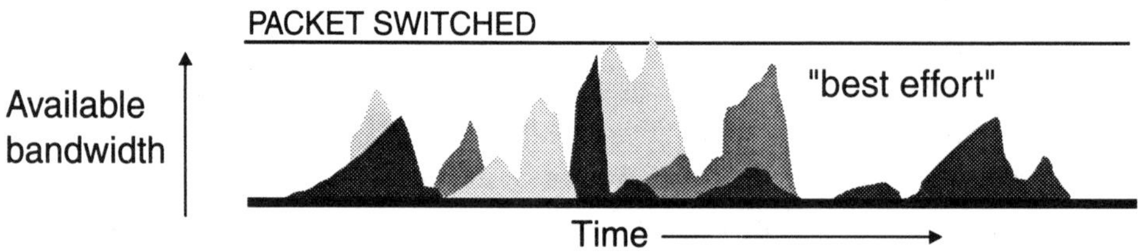

Packet-switched networks decompose all source media – whether audio, video, animation, text, software, or any of a number of other media types – into a series of digital packages, called *packets*. Each packet is routed through an open and shared network medium in a burst transmission that utilizes all available bandwidth to get each packet to its destination as soon as possible. When packets reach the destination computer, they are reassembled into the original source media.

This burst transmission mechanism is referred to as "best effort" delivery, and it effectively speeds packets along when there is little contention for network resources. However, all network-connected devices are free to send out as many packets as they need at any given time. Congestion can and does occur when too many packets concurrently pass through a particular part of the network. When this happens, packets are delayed and sometimes dropped (and must be retransmitted).

Packet-switched network transmission has been a good fit for the most popular phase 1 Internet applications: email messages, Web pages, database transactions, and instant messages. However, the best-effort delivery of packet switching offers no speed or delivery guarantees. In other words, most current packet-switched networks lack the Quality of Service (QoS) of circuit-switched networks. When congestion occurs, the lack of QoS can be annoying for those engaged in Web surfing and other phase 1 online activities; when it comes to rich media like audio and video, which is becoming central to phase 2 Internet interactions, congestion can disrupt the real-time flow of media, resulting in an unacceptable end-user experience.

Scaling packet networks for one-to-many media distribution – the core competency of the broadcast network – can also be problematic. This is especially prevalent with synchronous rich media transmissions, such as live streaming media Webcasts.

I introduced these challenges in the first book, and indicated then that significant funds were being invested within the converging telecommunications and computer networking industries to deal with the shortcomings of packet-switched networks. I was not alone when it came to high expectations for the future of the 'net.

Broadband Internet Everywhere?

During the Internet computing phase (phase 1) of digital convergence, the growth in the use of the Internet as a platform for human and organizational communication and interaction was phenomenal. As a result, the Internet has become a centerpiece of digital convergence, and it is generally recognized as an innovative global communications platform that offers the following advantages when compared to existing circuit-switched phone networks and broadcast television and radio networks:

- Media integration: Internet technologies support the ability to combine and transmit text, numbers, animations, audio, video, and other media types in any combination.

- Enhanced interactivity: The Internet can support one-to-one, group, community-based, and one-to-many modes of interaction. Interactions can be either synchronous (videoconference, chat, or live Webcast) or asynchronous (video on demand, email). Interactions to support searching, transaction processing, and navigation can be creatively integrated into an Internet interface.
- Network efficiency and reach: The best-effort, shared network approach of packet switching ensures more efficient network utilization for most applications. Also, information can be transmitted across any type of interconnected physical network.

Many believe the Internet has the potential to become the basis for all of our communications needs – with broadband Internet eventually knitting together a wider range of devices in a faster, more pervasive, and more functional network:

Internet-connected computers and PDAs

Internet phones

Internet television

Internet-connected automobiles

Internet kiosks

Internet vending machines

Internet appliances

Imagine a world where people just accept that "the network" (i.e., the next-generation Internet) is everywhere, interconnecting almost everything and everyone in the developed world. Legacy phone and broadcast networks would no longer be needed. Wireless networks would extend connectivity to an increasingly mobile population. Embedded processors and network connections – in dishwashers, home heating systems, autos, and even our clothes – would create an environment that I honestly still have trouble fathoming.

I remember in the 1980s and early 1990s when people used to talk about the Information Superhighway, which was then a dream – a dream of a high-speed network that offered ubiquitous, high-speed digital connectivity. No one knew who would build it. Would it be the phone companies, or the cable television providers, or some large long distance carriers? Or some combination?

While politicians, business executives, journalists, and others were debating the Information Superhighway, a community of geeks and hackers created the Internet, which rather quickly assumed a position as centerpiece of our digital communications consciousness.

The Internet evolved over the past few decades through a grassroots collaboration. It continues to evolve with new efforts focused on overcoming the technical shortcomings that prevent it from enabling innovative new services and eventually replacing legacy broadcast and circuit-switched networks. So, is the Internet is the Information Superhighway? And will we see broadband Internet everywhere? Many believe that this is indeed a sure bet. While I used to believe that myself, now I'm not so sure – at least not when it comes to the open, public Internet. Let's quickly review the past evolution of the Internet before sizing up the technological and industry drivers and challenges of this relatively new platform for human communication.

The Evolution of the Internet

The Department of Defense and academic researchers who originated the Internet were focused on creating a fault-tolerant communications platform based on a seamless network of interconnected networks: an internet. The Transmission Control Protocol and Internet Protocol (TCP/IP) that came out of these early efforts are intermediate-layer standard formats and procedures for transmitting data in packets of information across a heterogeneous mix of interconnected physical networks. The global public internet that eventually emerged with the broad deployment of these protocols is the Internet with a capital "I." The creators of the original Internet probably never fully envisioned the potential for this type of network or anticipated the breadth of applications that would ride on top of today's public Internet.

Initially, the development, standardization, and deployment of Internet protocols were voluntary and informal. While this worked remarkably well, there were some coordination issues, and the Internet Engineering Task Force (IETF) was eventually established to provide a structure for the Internet standards process. The IETF has provided a focal point and a set of formalized organizing principles for the collaborative, community-based work that has continued to churn out Internet innovations (the World Wide Web Consortium, or W3C, plays a similar role in coordinating Web standards).

The Internet was first widely adopted within the academic and research communities. During this time, in the 1980s, it became available at Syracuse University, and I recall how crude command-line tools such as Telnet and FTP became popular among only a small number of faculty, staff, and students. Email was popular with a larger number of users, but there were still many who eschewed this communications platform, owing to its crude interfaces and user unfriendliness. Then, when the Mosaic Web browser was introduced in the early 1990s – enabling a graphical interface to hyperlinked pages of text and images – its instant popularity launched the first phase of the digital convergence era.

My first article for *Network Computing* magazine was a 1996 comparative review of the Internet access capabilities offered by online community providers Prodigy, CompuServe, and America Online. This was about the time when Internet access was really taking off. These companies offered proprietary chat and bulletin board services, and they were beginning to offer limited Internet access to supplement their online services.* Through the late 1990s, tens of millions of consumers gained access to the Internet and Web via modem connections from local Internet Service Providers; regional and national Internet Service Providers such as MindSpring and EarthLink; and proprietary online service providers such as America Online, which, with its easy-to-use interface, became America's choice for those new to the Internet. During this period the corporate sector also saw widespread deployment of Internet technology.

In the early days most mainstream computer networking and telecommunications providers didn't take the Internet very seriously. Corporate computer networks were running packet-switched networks using proprietary protocols and services (e.g., Novell's IPX/SPX and Microsoft's Windows networking). And while broadcasters, cable providers, telecommunications carriers, and local phone companies recognized the importance of standards – organizations such as the International Telecommunications Union (ITU) and the National Television Systems Committee (NTSC) had established standards for these industries – they hadn't yet embraced the power and flexibility of packet-switched networks.

As more and more users jumped on to surf the Web and email their colleagues, business partners, friends, and families, and more and more companies began to understand the flexibility and value of the Internet, network engineers and computer geeks alike began to think and dream of ways this infrastructure could be fitted for a wider variety of applications.

This thinking drove extensions to the Web and the Internet that increased interactivity, supported transaction processing, and stimulated introduction of new rich media capabilities such as Internet telephony,

* AOL eventually acquired CompuServe on its way to becoming the number-one Internet Service Provider on the planet. Prodigy is still kicking, but it never mounted a serious challenge to AOL, and in November 2001 it was acquired by SBC Communications, one of the regional Bell operating companies discussed in the next chapter.

videoconferencing, and streaming media. The open and shared Internet was beginning to grow up, and it wasn't long before many of us began to believe that it was destined to become the Information Superhighway that had been talked about for so long. Broadband Internet everywhere became a reasonable expectation for this evolving, expanding – and increasingly general-purpose – communications infrastructure. It was easy to imagine a future Internet with incredible potential, and the game was on to extend and enhance the Internet to support all manner of human and organizational interaction, collaboration, communication, and information sharing.

Unfortunately, the Internet infrastructure was not able to adequately support some innovative new services and interactions. While there were high hopes, it was also clear that some crucial technical challenges had to be addressed for the promise to become a reality. Unfortunately, many of these issues still have not been adequately addressed, and I now believe that they may not be for some time to come – at least not in the context of the public Internet.

Daunting Technical Challenges

Bandwidth

As you may recall from the first book, the connection between a consumer premises and an Internet Service Provider (or any other telecom provider) is called the *last mile* or *local loop*. Substantial bandwidth is necessary in this crucial connection to ensure speedy transmission of high-resolution digital images, quick downloads of audio files in the compressed MP3 format, or smooth-playing streaming media feeds. Unfortunately, many consumers and small businesses continue to access the 'net through narrowband dialup connections at well under fifty thousand bits per second.*

The bandwidth solution for the local loop is a broadband connection – most often via cable modem connection to a cable provider such as Comcast or Time Warner, or a Digital Subscriber Line (DSL) connection to a phone company network.

Bandwidth is not only an issue in the last mile. With the open and shared architecture of the 'net there are other network components that also become bandwidth-constrained. For example: a crucial backbone inter-network router may become overtaxed when a sudden and unexpected surge in Internet traffic hits with a major news event; or a local Internet Service Provider may have a link to a backbone provider that works fine until a unique and compelling resource is deployed by a local ISP customer, and then Internet "word-of-mouth" traffic swamps the ISP's connection. While increasing bandwidth at these bottlenecks provides a quick fix, it is impossible to predict where bandwidth shortfalls will occur, and it seems there is a never-ending set of scenarios that cause bandwidth bottlenecks across the public Internet.

Let's focus in on two specific bandwidth-related Internet performance issues that keep the Internet from mounting a real challenge to legacy broadcast and circuit-switched networks.

1. Quality of Service

Real-time transmission of rich media or time-oriented text or numeric information (e.g., stock prices) requires a reliable transmission rate – i.e., Quality of Service (QoS) – between source and destination. QoS is a defining feature of the phone network, and it is something that Internet researchers and computer and network engineers have been pursuing for several years. Unfortunately, QoS is still not a reliable feature of the public Internet, a fact which has limited the quality, reliability, and ac-

* Effective throughput on a 56 Kbps modem connection is somewhere in the neighborhood of 33 Kbps.

ceptance of applications that incorporate real-time multi-media transmissions.

Among alternative Internet QoS mechanisms for the Internet are the following:

- *Over-provisioning:* This is the application of aggregate and ever-cheaper bandwidth to effectively raise the upper limit on transmission capacity to a point that contention and congestion never occur, even through peak network traffic spikes. Over-provisioning is not a total network-wide solution, but it is an effective strategy for some network segments, such as the Local Area Network (LAN), where it is possible to cost-effectively deploy enough bandwidth to prevent bottlenecks.
- *Resource Reservation Setup Protocol (RSVP):* This Internet protocol allows for an end-to-end reservation of bandwidth across the Internet for a particular transmission (e.g., a real-time audio or video stream). Unfortunately, it hasn't been feasible to implement RSVP across the public Internet – given the widely decentralized control over the many routers and other devices across the Internet that must support the protocol, and the complexities of billing and pricing for this type of fine-grained QoS approach on a totally decentralized network.
- *Multi-Protocol Label Switching (MPLS):* The MPLS standards effort has resulted in technology that can effectively tag different types of network traffic so that network administrators can program network devices to handle different types based on network "policies." For example, mission-critical corporate transactions or Internet video streams may be given priority over basic Web and email traffic. This is a preferred solution for many enterprise and service provider networks.
- *Caching:* While technically not a network QoS technology per se, propagating content to the edges of the network in local

caches enables local retrieval of content, which obviates the need for transmission across the public network, reducing the potential for network congestion and delay and significantly improving the prospects for a quality experience for the end user. Caching has given rise to the Content Delivery Network, which uses caching technologies to deliver quality Web and streaming media experiences (I introduced the CDN architecture in chapter 11).

In the foreseeable future it appears that Internet QoS will be limited to specific enterprise or ISP networks where financial models can support prioritization of specific types of traffic or caching of mission-critical content. These financial models do not translate well to the public 'net, given a strong resistance to usage-based pricing and the complexity of administering fees across a global decentralized network. The unfortunate result is that we are unlikely to see (at least anytime soon) the sort of creator-collaborator–centric, freewheeling and high-quality media access on the public 'net that many of us had dreamed of when the Internet emerged several years ago.

2. Scalability

The Internet excels at supporting millions of concurrent users. Part of this success is due to the fact that the typical aggregate demand is characterized by asynchronous access to an incredibly wide set of Internet-accessible resources. However, when many users all decide to access the same resource at the same time, bandwidth limitations often surface. As the horrible events of September 11, 2001, unfolded, news-oriented Web sites and ISP networks were swamped. Millions of 'net-connected individuals sought the latest in news and video, and access to mainstream Internet news sources degraded terribly. Television and radio, proven technologies for scaling to serve many viewers and listeners, provided the best window into the events of that fateful day.

Scaling up to synchronously serve content and interactions in real time to large numbers of users in a reliable, cost-effective manner is a huge challenge. When many Internet users try to concurrently access a single resource, a separate transmission is required for each and every user. This is the root cause of the scalability challenge.

There are a variety of solutions designed to enable scalable access to popular Web resources and rich media events:

- *Multicasting:* This is a technical solution for efficiently propagating real-time transmissions across the Internet infrastructure for many users, conserving bandwidth and ensuring a quality experience for end users. This is particularly promising for some broadcast-style events. Unfortunately, an inability to consistently implement multicasting capability across all Internet routers renders this solution ineffective for public Internet transmissions. To this point, multicasting offers a reliable delivery option only within a managed environment such as a single ISP or enterprise network.

- *Load balancing:* Responding to sudden spikes in Web traffic will increasingly be handled by load balancing technology (from F5 Networks, RADirect, and others) that can dynamically allocate bandwidth across service provider networks. This solution is appropriate for applications targeted for many concurrent users.

- *Caching:* Introduced above as a technology that can enable QoS, caching can also be used to scale Webcasts or popular Web content to serve many concurrent users.

Various Internet Service Providers will continue to develop and implement scaling mechanisms to gain competitive advantage, primarily when aiming offerings at well-funded enterprises. However, at this juncture there is no completely reliable, efficient, and cost-effective solution for one-to-many scaling across the public Internet for all scenarios.

Security

While the openness of the Internet is one of the powerful characteristics of this new communications platform, it has also created an ominous dark side as unscrupulous creator-collaborators wreak havoc on the information and processing capabilities of others.

When I wrote the first book, things like email viruses and Web-server security intrusions were largely an inconvenience that only occasionally resulted in serious data loss or system disruption for most users and organizations. But as the growth and use of the Internet has continued to empower those who work in the dark side of the 'net, we've seen a marked increase in the number and severity of security events, with the costs of these events increasing in terms of time and resources.

- Email-based virus attacks are running rampant.
- Automated probes for potentially unprotected desktops and servers are now a fact of life.
- Peer-to-peer file sharing programs have provided a new platform for transmission of worm and Trojan horse programs.
- Denial-of-service attacks, where many broadband-connected PCs are commandeered to launch a packet attack on an unsuspecting network, have become commonplace.
- Internet-connected devices which feature embedded Web servers, such as printers, are being targeted for disruptive intrusions.
- There are now growing concerns over PDA and cell phone viruses.*

---

* Patricia Cardoza, "Block PDA Viruses," ZDNet Tech Update, April 25, 2002 (http://techupdate.zdnet.com/techupdate/stories/main/0,14179,2862764,00.html).

Lynne Burke, "About That Cell Phone Virus," *Wired News*, June 8, 2000 (http://www.wired.com/news/politics/0,1283,36843,00.html).

The increase in Internet security intrusions has cast doubt on the suitability of this infrastructure for some of our most important services and interactions. If we are to eventually see broadband Internet everywhere, there must be some assurance that our information and our technologies can remain secure. This clearly is in doubt.

During a time of increased terrorism in our society, and of decreasing trust among the global citizenry, the suitability of the Internet for doing business and developing reliable, high-quality services has been increasingly called into question. When the U.S. government became serious about implementing a separate Internet network, called GovNet, it became clear to me that the mounting security problems of the 'net had reached a significant, negative threshold.*

A variety of technologies and mechanisms are being put into place to combat new threats to our use of the Internet. Secure corporate connections across the public Internet are being proactively implemented using Virtual Private Network (VPN) technology. Network intrusion detection tools, undercover efforts to infiltrate hacker rings, and advanced antivirus programs help us to identify and combat security threats. Unfortunately, it seems we're still too often reacting to new threats after they've already struck.

Metcalfe's Law: The Downside

The decentralized Internet has been created through voluntary support for standard protocols and services and a collaborative assemblage of network interconnections. Any organizational network or ISP anywhere can interconnect and provide new communities with Internet con-

---

* Sandra Swanson and Steve Konicki, "Internet Too Unsafe for Feds," *InformationWeek*, October 11, 2001 (http://www.informationweek.com/story/showArticle.jhtml?articleID=6507532).

nectivity. Any individual is free to connect to the 'net using any of a number of ISP edge networks. The universal interconnection that this provides is clearly one of the most powerful advantages of the Internet. It is a beautiful thing.

Robert Metcalfe, the man who invented Ethernet, came up with an interesting rule of thumb, now known as Metcalfe's Law. It captures the notion that the usefulness of any network is tied to the number of individuals and the breadth of resources connected. Specifically, he said that the value of a network can be expressed as n squared, where n is the number of nodes connected to the network. The Internet has evolved into communications powerhouse largely because of open and universal access and the impact of Metcalfe's Law.

Unfortunately, the open, decentralized architecture and collaborative evolution of the Internet also raises some serious issues and concerns as the number of endpoints grows to include much of humanity. Many of these issues will never be fully addressed across the entire public Internet, at least anytime soon. This opens the door for proprietary solutions that address these issues on single-operator networks – which plays into the hands of large profit-driven communications providers who seek desperately to corral and control market share. We will explore this further in the next chapter.

13. In the Hands of Network Giants

> "The lesson we should have learned from the failure of the 1996 Telecom Act is that the interests of major telecommunications companies and average American consumers are not the same."
>
> Senator John McCain
> From the floor of the Senate, May 1, 2002*

During phase 1, the Internet leveraged existing telecommunications infrastructure and technology to introduce innovative new capabilities to millions of people worldwide. Throughout this phase there were plenty of winners, and few losers. Phone companies sold more phone lines to consumers without an investment in Internet technology at the endpoints, since ISPs and consumers provided the modems for connecting. Long-haul telecommunications carriers were able to markedly increase their inventory of leased digital telecommunications lines to ISPs and corporations. Cisco Systems emerged as primary provider of the Internet router infrastructure, becoming a major new force in the communications technology marketplace in the process.

The Internet has now moved well beyond infancy into adolescence, and in phase 2 we've begun to see some warning signs that indicate a troubled teen. Even so, the value of Internet technologies as a platform for innovative communication services has been recognized by all major communications providers.

Unfortunately, it seems there is sometimes more hype than real action from these giant players – as the gap between what the technology could enable and what is rolled out to consumers is often significant. The bold vision for a high-performance open, public Internet everywhere is

* You can find Senator McCain's complete statement on his Web site (http://mccain.senate.gov/index.cfm?fuseaction=Newscenter.ViewPressRelease&Content_id=447).

challenged not only by the technical issues raised in the last chapter, but also by the market dynamics, monopolistic practices, and regulatory constraints that are part and parcel of today's increasingly consolidated digital communications marketplace. The zeal for progress has been tempered by a bottom-line conservatism – borne not only from technical challenges, but also from corporate disdain for relinquishing market share and an abhorrence for the killing of any sacred cash cows.

In the United States, the federal government has been active in regulating telecommunications businesses to (hopefully) stimulate competition, ensure near-universal access to services, and limit abuses such as price gouging.

- The Congress creates legislation that the FCC interprets in its rulemaking activities, as was the case with the Telecommunications Act of 1996.
- The Federal Communications Commission (FCC) is the bureau most often tasked with creating and enforcing rules that dictate the behavior of telecommunications carriers.
- The Department of Justice and the courts get involved when disputes arise.
- Mergers of large communications companies are approved or denied by the Federal Trade Commission or the Department of Justice.

While the U.S. government has tried to enable competitive access to phone networks and stimulate the rollout of innovations such as broadband connections for consumers, it has often been an uphill battle.

Let's review the various types of telecommunications providers to help you understand the major players and what motivates them. In the process I'll touch on some of the present-day realities that are impacting the evolution and deployment of second-phase telecommunications technologies and capabilities. We'll wrap up the chapter by highlighting the needs of the creator-collaborator, which offer quite a contrast to the services and functions provided by many phase 2 communications providers.

The Players

Phone Companies

When I was a kid, I remember people talking about "the phone company" with reverence. When it came to interactive, electronic communication it was – like the "power company" in the realm of electrical service – the only game in town. At that time giant AT&T owned twenty-two Bell local phone system operators that provided the lion's share of the U.S. population with local phone service, and it provided the circuits used for most long distance calls.

As microwave and other new communications technologies emerged in the latter half of the last century, alternative carriers such as Sprint and MCI entered the long distance marketplace. AT&T resisted opening up local circuits to these new carriers, since it wanted to maintain its monopoly control, and this effectively disabled real competition in the long distance market.

In 1984, the U.S. Department of Justice filed an antitrust suit that resulted in the breakup of the AT&T monopoly; this breakup created seven regional Bell operating companies (RBOCs), each owning the Bell phone companies within a specific region (see Figure 1).*

* AT&T retained its right to offer long distance service and kept its telecommunications research and manufacturing operation, which eventually became Lucent Technologies.

Figure 1. The Original Seven Regional Bell Operating Companies

Ameritech	Midwest states
Bell Atlantic	Atlantic states south of New York
Bell South	Southern states
NYNEX	New York and New England
Pacific Telesis	California and Nevada
SBC Communications	Southwestern states
US West	Mountain and western states

While the breakup of AT&T did energize the long distance market – since RBOCs were anxious to provide local phone connections to competitive long distance carriers – there was still only limited local phone competition. New technologies had enabled some newcomers, called Competitive Local Exchange Carriers (CLECs), to enter the market, but they were limited in scope. CLEC services were focused primarily on business customers in metropolitan areas, leaving most U.S. consumers with no viable alternatives to the Incumbent Local Exchange Carrier (ILEC) and its pervasive local network.

Local telephone service remained largely under the control of the RBOCs, with the FCC and state public service commissions overseeing many local phone service regulations. This was not a good situation: restrictions on RBOCs kept them from providing long distance service; RBOCs couldn't provide new data services, such as Internet access; and long distance and cable providers were unable to offer local phone service in some regions. The monopolistic and regulatory quagmire was stifling competition and deployment of advanced communications services.

So once again the federal government moved to stimulate competition with the Telecommunications Act of 1996. The Act required RBOCs to provide local competitors with access to their networks and facilities; and if RBOCs could subsequently demonstrate that they had adequately opened their networks to competition, they would be allowed to compete in the long distance market. It also cleared the way for a variety of other providers, such as long distance companies, CLECs, and cable companies, to provide local phone service.

While the Telecommunications Act of 1996 seemed promising at the time, the overall impact of this legislation, in the estimation of many, came up short.

- With the way cleared for RBOCs to offer advanced services, they were relatively slow to roll out broadband DSL Internet service, one of the most innovative data services to come down the pike in decades.
- In many instances, RBOCs were less than cooperative in opening up local networks to competitive carriers.
- Many consumers have only recently seen local wireline phone service competition.

To gain entrance into long distance markets RBOCs were required by the FCC to demonstrate that they had opened their networks and enabled significant local phone service competition. It took until the end of 1999 before the first RBOC successfully submitted an application to the FCC to enter a long distance market. BellSouth was the last of the RBOCs to get its first approval in 2002.* As of May 2003, RBOC long distance service was available in 41 states (after many applications had been denied or withdrawn).**

Cable Companies

The cable industry was originally launched to provide expanded television viewing options for consumers in remote areas not adequately served by terrestrial broadcasts. From this humble beginning, the cable

* Michael Martin, "RBOCs on Long Distance Trek," *NetworkWorldFusion*, May 27, 2002 (http://www.nwfusion.com/news/2002/0527carrier.html).

** "Statistics of the Ling Distance Telecommunications Industry," Federal Communications Commission, Industry Analysis & Technology Division, Wireline Competition Bureau, May, 2003 (available at http://www.fcc.gov/wcb/stats).

industry has grown into a major force in the communications and entertainment industries.

Like the local phone company and your electric company, the local cable provider has a significant investment in wired infrastructure – an infrastructure that cannot be easily duplicated by a potential competitor. And like the local phone company, the cable company is typically a local monopoly, providing a single wireline option for cable television service. However, unlike phone companies, which must open up their networks to competitors as specified in the Telecommunications Act of 1996, cable providers to this point have not been required to open up their networks to third parties.

The cable network also differs from the phone network from a technical standpoint, as described in some detail in the first book and in chapter 12. To quickly review: phone networks are digital circuit-switched networks that utilize copper cabling and analog electrical signals in the last mile to carry voice conversations; cable networks are broadcast networks which use a head-end facility to aggregate video sources (mostly satellite-based) which they deliver across a hybrid fiber-optic/coaxial-cable network.

As the digital convergence phenomenon began to take shape, cable providers saw a great opportunity to provide additional services and to take market share from local phone companies as well as Internet Service Providers – and (hopefully) become the next-generation data network for the masses. While local phone companies dragged their feet on DSL, cable providers had a jump on the broadband Internet marketplace and they quickly became the clear leaders in deployment of broadband Internet services.*

* A Gartner Dataquest report, *U.S. Market Loves Broadband More Than Ever*, indicated that as of June 2002 cable modem had a 54% market share while DSL had reached a 34% market share, as summarized in: "U.S. Broadband Growth Steady," *CyberAtlas*, November 20, 2002 (http://cyberatlas.internet.com/markets/broadband/article/0,1323,10099_1545711,00.html).

Cable networks used to be strictly analog networks, providing a one-way broadcast of conventional and cable-only video programming services. Now most cable providers have upgraded networks with fiber-optic cabling to local neighborhoods and implemented bi-directional, digital interactive capabilities. The result is a very capable digital network infrastructure, with most cable operators now offering digital cable television services as well as high-speed Internet access. Voice over broadband cable appears to be next on the agenda.

During phase 2, cable operators will leverage the current control they exercise over television-based consumer entertainment and their lead in broadband Internet access. Video is one of the rich media types that I expect will be transformed in the next phase, and the cable providers recognize this. While cable companies have been relatively aggressive in rolling out broadband and other new services (e.g., they are now moving forward with voice over IP services), they will move slowly when it comes to opportunities based on open IP-based video content services – especially open-access streaming media services – given the potential disruption to conventional broadcast revenue streams.

Cable providers are also in the best position to deploy advanced services in set-top boxes that combine broadband Internet sharing and interactions with advanced video services. The television set-top box could be among the most important technologies of phase 2. Unfortunately, the cable industry is likely to offer relatively limited access and capabilities in any set-top boxes deployed in the near term, since advanced capabilities with open Internet access could very well lead to increased market fragmentation and contribute to reductions in conventional cable revenue streams. Set-top boxes are discussed in more detail in chapter 16.

Emerging Forces

The phase 1 growth in the use of the Internet, and the hype that surrounded that growth, stimulated the launch and rapid growth of a new wave of communications ventures. Many were aggressively funded, and they provided innovative new services and useful new infrastructure. Many new technology and service initiatives were also launched by conventional telecommunications providers. While many new entrants were decimated by the bursting of the dot.com bubble and the subsequent downturn in the telecom sector, the telecom sector is coming back and there are some compelling survivors. Here is a quick rundown on some of the alternative players in the telecommunications marketplace.

1. Internet Service Providers

ISPs run the gamut from those that operate state-of-the-art backbone networks and cater to the needs of smaller ISPs and large corporations, to consumer-oriented ISPs (such as NetZero and AOL), to smaller regional and local network providers (such as Dreamscape here in Syracuse).

ISPs are often categorized according to a set of tiers, with any definition inviting controversy from all quarters of the ISP marketplace. Suffice it so say the high-end tier-one providers are those ISPs that run extensive networks and sell bandwidth to other providers. At the low end, tier-three providers are small-scale providers who lack a national backbone and must connect to larger providers to provide end-to-end Internet connectivity. In between are the tier-two ISPs (these are typically regional providers).

The ISP market saw considerable consolidation and bankruptcy during the early 2000s. Among the bankruptcy victims were companies such as Applied Theory, a regional ISP which counted Syracuse University among its customers, and PSINet, a one-time ISP powerhouse whose name once adorned the football stadium of the NFL's Baltimore Ravens. Still, some compelling ISPs, such as Broadwing and Earthlink, remain.

The late 1990s saw a wave of rapidly expanding long-haul and Internet Service Providers with extensive national or global networks based on optical networking technologies and Internet protocols. Level 3, Global Crossing, Broadwing, Qwest, and Cable and Wireless are among this wave. Other high-speed providers, such as WorldCom (now MCI) and Sprint, have also invested in national and global IP networks. Unfortunately, many of these providers have also seen financial trouble, with Global Crossing and WorldCom among the bankruptcy victims. The survivors possess significant high-speed, IP-based network infrastructure assets.

Many ISPs offer services beyond the basic Internet connection – such as Web hosting, e-commerce services, and email hosting – to differentiate their services from other communications providers.

A new class of ISP, the Wireless ISP (or WISP), is emerging to provide subscriber-based broadband access using relatively inexpensive and easy-to-deploy wireless technology. WISPs provide a viable alternative to RBOC and cable company broadband in many cities, and entrepreneurs are launching WISPs in rural areas and other areas where wired broadband isn't an option. Wi-Fi technology enables these new providers to quickly enable broadband connectivity without the significant investments required to build a wired infrastructure.

2. Broadband Service Providers

The Federal Communications Commission identified this new class of communications service provider in 2001 in its annual report on competition in the video delivery sector.* The group includes smaller competitive cable network providers as well as Competitive Local Exchange Carriers (CLECs) that survived the telecom downturn. A number of these providers maintain fairly extensive state-of-the-art networks capable of providing voice, video, and data services.

CLECs initially focused on alternative voice service for corporate customers in large metropolitan areas. During the late 1990s, many CLECs pioneered new services and technologies (e.g., in many cities CLECs were the first to provide DSL services, beating RBOCs to the

* From the Federal Communications Commission, *Annual Assessment of the Status of Competition in the Market for Delivery of Video Programming, Tenth Annual Report*, January 5, 2004.

punch). The downturn in the tech sector at the turn of the century, combined with the alleged uncooperativeness of RBOCs, which were required to open up their facilities to competitive carriers, forced many CLECs, including Winstar Communications, Covad Communications, Teligent, and Rhythms Net Connections, into bankruptcy proceedings.* Several surviving CLECs, such as Winstar (which was purchased by IDT) and XO, are now offering compelling high-speed communications solutions based on fixed wireless and other new technologies, mostly aimed at business customers.

Other broadband service providers include cable "overbuilders" (companies that have built out a second competing cable infrastructure in some cities), such as RCN and WideOpenWest (WOW), which have gained significant market share and are offering a variety of advanced cable services.

4. Cellular Network Providers

Cellular providers have emerged as a major force in the communications marketplace, and most are gearing up to provide high-speed data services to mobile users (this is covered in chapter 17). While most of the major players in the U.S are owned by RBOCs and the major long distance providers – this includes AT&T Wireless/Cingular (SBC and BellSouth jointly own Cingular, which recently acquired AT&T Wireless), Verizon Wireless (a joint venture of Verizon and Vodafone), and Sprint PCS – there are a few other players in the United States that offer innovative technologies and approaches to gaining market share in this marketplace. These include Nextel, T-Mobile (formerly VoiceStream; now owned by Deutsche Telekom), Cricket Wireless (a company of Qualcomm spinoff Leap Wireless, which has positioned its service as an al-

* David M. Ewalt, "Covad Plans Bankrupty, Stock Swap," *InformationWeek*, August 7, 2001 (http://www.informationweek.com/story/showArticle.jhtml?articleID=6506390).

ternative to conventional wired phone service), and several regional providers.

4. Others

A host of other companies offer wired and wireless services that compete with network services offered by incumbent phone companies and/or cable providers.

- Satellite providers EchoStar and DirecTV compete for television customers with cable providers. DirecTV also operates a satellite broadband Internet service, called DirecWay, and EchoStar has launched additional satellites to offer advanced network services including high-speed Internet access.*
- Next-generation satellite providers such as Inmarsat and Mobile Satellite Ventures are now offering voice and data services, and some have approached the FCC in an effort to gain access to urban markets and to deploy terrestrial wireless components in urban areas and buildings where satellite signals are blocked.**
- A variety of network providers have launched Wi-Fi networks in open public areas such as airports and commercial districts, providing broadband connectivity for consumers and business pros on the go. Wi-Fi aggregators enable a single subscription that enables access across many of these networks. Aggregators such

* Federal Communications Commission, *Annual Assessment of the Status of Competition in the Market for Delivery of Video Programming, Tenth Annual Report*, January 5, 2004.

** Federal Communications Commission, *Annual Report and Analysis of Competitive Market Conditions With Respect to Commercial Mobile Services, Eighth Report*, June 26, 2003.

as Boingo, GRIC, and iPass provide uniform access to Wi-Fi networks scattered throughout world.*

- Power companies are exploring the possibility, with FCC support,** of using ubiquitous power networks for new broadband services (this is referred to as Broadband over Power Line, or BPL).

Competition, Consolidation, and Regulation

As the pace of telecom technology development has quickened, the competition among the various sectors of the communications industry has increased, and the major players in each sector have sought to consolidate. This consolidation – which enhances political influence, extends available resources and infrastructure, and increases market clout – is said by some to strengthen the few remaining providers in all sectors for an increase in intermodal competition (i.e., competition among providers using alternative technologies and networks). Intermodal competition is created by convergence as satellite, telephony, wireless, and cable television providers move to end-to-end digital transmission.

While some believe the competition between the modes will be strengthened with reduced regulation and government controls on industry consolidation, many others believe that deregulation and consolida-

* As of January 24, 2004, Boingo had over 5,000 locations under contract and over 2,100 live worldwide. From the Boingo Web site (http://boingo.com/services.html).

** Roy Mark, "Broadband Over Power Lines: The FCC Plugs In," *InternetNews*, April 23, 2003 (http://www.internetnews.com/bus-news/article.php/2195621).

tion will simply increase monopoly control over existing and emerging markets and may in fact constrain competition and innovation.*

Regardless of the debate, the momentum toward a digital communications marketplace with a smaller number of larger companies is undeniable.

RBOCs have moved to strengthen any remaining monopoly power by merging with other carriers and with other telecommunications companies to create a new set of telecom powerhouses:

- Verizon – a combination of NYNEX and Bell Atlantic, along with Verizon Wireless
- SBC – Pacific Telesis, Ameritech, and the former SBC combined into one; SBC also owns sixty percent of Cingular, now the largest wireless provider in the U.S.
- Qwest – Qwest, a communications provider with an advanced IP-based fiber-optic network, acquired former RBOC U.S. West in the 1990s
- BellSouth – this southern RBOC owns forty percent of Cingular

Cable providers have also been consolidating, with larger providers buying smaller metropolitan cable operators over the past several years. The result is a small number of very large operators who control most of the U.S. market: AOL Time Warner, Charter Communications, Cox

* Vinton Cerf offers an intriguing opinion on intermodal competition in a May 20, 2002, letter to Donald Evans, Secretary of the U.S. Department of Commerce, and Michael Powell, Chairman of the U.S. Federal Communications Commission (http://global.mci.com/us/enterprise/insight/cerfs_up/issues/broadband_letter.xml).

Communications, Adelphia, Cablevision, and Comcast (which merged with AT&T Broadband to become the market leader).*

We've also seen consolidation among long-haul and backbone providers. Sprint and MCI got into the long distance game many years ago, and a variety of other regional and national long distance providers entered the market over the years. One of these was LDDS (Long Distance Discount Service), a small long distance reseller that eventually grew through acquisitions to become powerhouse WorldCom, which in 1997 acquired MCI. WorldCom also acquired MFS Communications, a large CLEC which had itself acquired UUNet, a large Internet Service Provider. In the summer of 2001, WorldCom attempted to acquire Sprint, a move blocked by government regulators. WorldCom's bubble burst in a flameout of corporate greed and mismanagement, and it filed for Chapter 11 bankruptcy protection in July of 2002. The lowered debt burden of the restructured company, once again known as MCI, may make it a potent telecom competitor in phase 2.

All the major long distance companies have evolved and grown, and they now provide long distance, leased-line data, and high-speed Internet services for large corporations at minimum. AT&T's Worldnet service has also become a leading consumer ISP, and Sprint and AT&T developed two of the largest nationwide cellular networks (as of this writing AT&T's cellular network is in the process of being acquired by Cingular, but AT&T has indicated that it will likely re-enter the cellular marketplace once the acquisition is completed).**

---

* The $47.5 billion merger of Comcast and AT&T Broadband, completed in 2002, combined the two cable systems to serve 21 million video customers in 41 states along with 6.3 million digital video customers, 3.3 million high-speed data customers, and 1.3 million cable phone subscribers. From the Comcast Web site (http://www.cmcsk.com/phoenix.zhtml?c=147565&p=irol-events).

** Bruce Meyerson, "AT&T Sees Better Times Ahead," *InformationWeek*, February 25, 2004 (http://www.informationweek.com/story/showArticle.jhtml?articleID=18200612).

It doesn't end there. America Online, the largest consumer-oriented ISP, merged with Time Warner, the media giant that is also among the leaders in the broadband cable modem market. DirecTV and EchoStar (which operates Dish Network) tried to merge, claiming that they needed to join forces to effectively compete with cable providers. The Federal Communications and the Department of Justice both rejected the merger, due to concerns over reduced competition in pay television.* SBC has also pursued the possibility of acquiring DirecTV.**

These giants all claim that consolidation through acquisition and merger is necessary to remain competitive, since innovative new services require considerable investments that only a large provider can underwrite. They also claim that previously distinct categories of providers are all in competition now as the move to converged high-speed digital networks enables communications providers to increase the breadth and depth of their offerings.

Of course these large companies are also focused on the bottom line, and they loathe the disruptions and the potential for declining revenues that often accompany innovative offerings. In the telecom industry there is a history and culture around monopolistic practices and a resistance by incumbent carriers to timely deployment of new technologies and services (e.g., RBOCs were allegedly slow to roll out DSL services partly because of the fear of losing significant revenue from much more expensive T1 leased-line services).

The debate over whether cable and RBOC networks must be open or closed to competition is another point of contention – with conflicting opinions on how the relative openness of public networks will constrain or enhance competition and innovation.

* "Satellite TV Merger Nixed," *Wired News*, December 10, 2002 (http://www.wired.com/news/business/0,1367,56792,00.html).

** Penelope Patsuris, "SBC Reportedly Goes After DirecTV," Forbes.com, February 7, 2003, (http://forbes.com/2003/02/07/cx_pp_0207sbc.html)

The FCC has developed different rules for different types of communication services. From a regulatory standpoint, local phone networks operate what are considered "telecommunications" networks; cable operators offer an infrastructure with a broadcast service that is considered "cable service"; and cable-modem–based Internet access is categorized by the FCC as "information service." While phone companies have been forced to open up their telecommunications networks, cable providers have not been required to open their cable services or information services. This differential treatment is a regulatory underpinning that helped stimulate cable operators to aggressively roll out new digital interactive services without the fear of having to eventually open up these services to competitors. Phone companies have been reluctant to invest owing to concerns that the resulting infrastructure would be made available to competitors at very low prices.

Proponents of reducing or eliminating open access requirements point to the evolving offerings of long-haul, RBOC, cable, wireless, and satellite providers as evidence of intense competition within and across telecommunications marketplaces. Opponents argue that further deregulation would increase the power of monopoly providers, further eroding the competitive landscape.

The current administration of the FCC appears to favor a move away from forced open-access requirements.

- A 2003 court challenge to the FCC classification of cable-modem service resulted in a 9th Circuit Court ruling that cable-modem service is part telecommunications service and part information service,* opening the possibility that cable providers may also be required to open their networks. The FCC is appealing this ruling.

* "Court Reverses FCC Cable Ruling," *Wired News*, October 7, 2003 (http://www.wired.com/news/business/0,1367,60726,00.html); as of this writing the FCC is planning to appeal this ruling.

- In February 2003 the FCC excused the RBOCs from opening their broadband networks to competitors.*

Clearly, the evolution of the communications industry and its offerings is driven by a very wide variety of factors. Unfortunately, as giant communications companies continue to consolidate and government officials react to intense corporate lobbying efforts, the most compelling and useful features of digital communications services – openness, flexibility, innovation, and decreasing prices – are too often lost in the shuffle.

What Do Creator-Collaborators Want?

For most creator-collaborators, the key characteristics of attractive phase 2 network access services will not be content services, program guides, or even ease of use. We favor network services that will allow us to push the envelope on creative uses of computing, digital media, and communications technology, and allow us to further our personal and professional agendas.

Here then, from my creator-collaborator standpoint, are what I believe are the most important aspects of prospective phase 2 digital connectivity options.

- *Openness:* Unfortunately, large telecommunications providers are increasingly interested in deploying closed network services that they will claim are necessary to ensure quality and perform-

* "FCC Adopts New Rules for Network Unbundling Obligations of Incumbent Local Exchange Carriers," February 20, 2003 (http://hraunfoss.fcc.gov/edocs_public/attachmatch/DOC-231344A1.pdf).

Elisa Batista, "Few Cheers for New FCC Rules," *Wired News*, February 21, 2003 (http://www.wired.com/news/business/0,1367,57755,00.html).

ance, but which are also designed to ensure clear-cut markets. Walled gardens and closed content libraries (e.g., when cable providers or 3G-based cell providers roll out advanced video services, but restrict access to fee-based content developed by mass media partners) are examples of proprietary approaches that most creator-collaborators loathe. Creator-collaborators are most often focused on the one-to-one, small group, and community-of-interest interactions that are enabled by openness – as exemplified by the phase 1 Internet.

- *Performance:* Bandwidth is crucial to the creator-collaborator. Current broadband solutions are based on asymmetric access networks, assuming that passive consumers would spend more time downloading Web and streaming content, without a need to upload content and services. Broadband providers have been surprised by the need for a fast uplink and are now scrambling to provide additional bandwidth in the upstream direction. Creator-collaborators often share a broadband pipe with multiple PCs – and perhaps other devices as well. New tiered pricing for broadband offerings capped at different transmission rates is not a solution creator-collaborators welcome. Creator-collaborators want everyone to have a big fat pipe, enabling reliable very-high-speed connectivity to all our colleagues, friends, and family members and to all server and peer-to-peer computing and information resources.
- *Price:* If you're like me, you'd like a big fat digital pipe that you pay for with a flat fee but that you can use for multiple purposes – including telephony, Internet access, and entertainment. Many creator-collaborators feel nickel and dimed. There are fees for cable TV service; fees for cable-modem service; fees for landline phone service; and fees for wireless cell phone use. You're talking well over a hundred bucks by the time you're done. Creator-collaborators want open, high-bandwidth, quality connections

that are offered at a reasonable price, and at a fixed rate. I have difficulty understanding why my cable bill continues to rise and my phone bill remains constant in the face of decades of technological advances that can provide higher-capacity, less-expensive communications solutions.* While I've never spent the time to track the trail of dollars and cents, it seems fairly obvious to me that something is wrong with this picture!

- *Mobility:* This is important for many creator-collaborators, especially those of us who are looking for a more mobile lifestyle that keeps us connected to our clients, colleagues, business partners, or customers. Still, for many creator-collaborators, a higher priority is an affordable, open, high-speed Internet connection into the home and/or workplace that enables creative PC access to much more functional interactions and resources. Interestingly, consumers may well place a greater emphasis on mobility, because of its convenience and because they often fail to grasp and embrace the empowering potential of convergence technologies.

Imagine . . .

A colleague who read this chapter said he felt that in the above section I should specify just how much bandwidth creator-collaborators need. Is one million bits per second enough? Or maybe five, ten, or twenty million? I mulled this over and decided I'd be doing the creator-collaborator movement a disservice by assigning a number. It's just not that simple.

---

* Between 1993 and 2003 cable rates increased 53.1% while the Consumer Price Index increased 25.5%. Cable operators attributed rising costs to increased programming costs and higher labor costs. From the Federal Communications Commission, *Annual Assessment of the Status of Competition in the Market for Delivery of Video Programming, Tenth Annual Report*, January 5, 2004.

The value of increased Internet bandwidth is the way in which it stimulates creator-collaborators to invent new ways to communicate, interact, and share information. I believe that the more bandwidth that is available, the more aggressive and more fruitful these efforts – and the more stunning the results. Digital convergence phase 1 has demonstrated the concept. Now, just try to imagine what might be possible with a LOT more bandwidth.

Not too long ago I saw one of the cable network CEOs (I think it was Comcast CEO Brian Roberts, but I can't recall for sure) discussing the success of broadband Internet on CNBS or CNNfn. He suggested that his company is only using a single analog channel's worth of bandwidth for broadband Internet, and he mused over the possibility of using more than one channel.

I had never looked into cable television spectrum usage in any detail, and I must admit I was surprised that such a powerful service could be delivered with such a small slice of the transmission capacity of the cable wire coming into my home.

I did some research* and found that each analog channel uses 6 MHz of spectrum, and that broadband Internet access via cable modem typically uses the equivalent of a single 6-MHz "channel" for downstream connectivity, while one upstream channel uses up to 3.2 MHz of radio frequency bandwidth. This arrangement can accommodate up to forty million bits per second of data traffic in the downstream direction, and up to ten megabits per second for upstream traffic with first generation (DOCSIS 1.0 and 1.1) cable-modem technology. This bandwidth is typically shared with anywhere from a few hundred to a couple of thousand homes in a single given neighborhood.

Given that today's hybrid fiber-coaxial cable network has the capacity for over 100 channels, it is clear that only a very small portion of to-

* "Cable Modem/DOCSIS Project Primer," last updated on October 27, 2003, from the CableLabs Web site (http://www.cablemodem.com/primer).

day's cable television infrastructure is devoted to broadband Internet access.

Imagine if cable companies would convert just ten percent of their analog channels – as opposed to just one or two – to open Internet access for each neighborhood. The result would be super-fast Internet access that would stimulate a real surge in interactive and on-demand rich media use.* Creator-collaborators would lead the charge in utilizing available capacity, and I honestly believe that profound new interactions and resources would emerge. Computing and digital media technologies have evolved very nicely over the past several years and are ready for prime time when it comes to new emerging forms of digitally-enabled human interaction. What we need to seal the deal – from a technology standpoint – is more bandwidth.

I would gladly give up a dozen or so analog cable channels in exchange for a super-fast Internet access and the ability to explore new modes of rich-media access and interaction (please, take all the home shopping channels asap!).

So, do you think this will happen anytime soon? In my opinion it's not likely – even if it were inexpensive and easy to do, which is probably not the case. Cable and entertainment providers have too much to lose to go ahead and open the floodgates to an open platform that supports high-quality content and interactions.

With HDTV finally gaining momentum, any move to free up transmission capacity will be eaten up by the move to the more bandwidth-intensive high-definition format. Cable providers have a keen incentive to ensure that broadband access to the public Internet continues to deliver inferior-quality video media in comparison to their advertising-supported closed broadcast delivery mechanism.

---

* Of course I'm aware that other portions of cable provider networks and the Internet would be affected by this, but let's not get bogged down in those details here – the idea is to think out of the box.

Still, I continue to hope that before my lifetime has ended we see a day when high-speed open access to rich media resources and interactions from across the face of the planet will overcome the closed paradigm of broadcast, cable, and satellite television.

Phase 2 Internet evolution has moved squarely into the domain of large telecommunications and cable companies, a development that may well constrain the evolution of the Internet infrastructure across some important dimensions. Perhaps some disruptive innovation will again obviate the influence of the large communications carriers. The golden rule of technology feeds the grassroots geek creator-collaborator mentality, and it could set the stage for some new technology (e.g., a new wireless technology like ultra-wideband, which offers gigabit wireless connections over short distances,* or WiMAX, which can support 74-Mbps wireless connections over tens of miles)** that can re-energize the grassroots evolution of the open public Internet.

In the next section we'll explore the evolution of the services, interactions, and tools that are emerging to provide exciting new capabilities that can enhance the activities and the lives of creator-collaborators.

---

* Peter Meade, "Ultra-Wideband: New Ultra-Connectivity for the Enterprise?" *Networking Pipeline*, December 3, 2003 (http://www.networkingpipeline.com/showArticle.jhtml;jsessionid=GT2SRT14DXHE0QSNDBCSKHQ?articleId=16401583&pgno=2).

** Rick Merrit, WiMax Forum Targets First Systems by End of Year," *Mobile Pipeline*, January 21, 2004 (http://www.mobilepipeline.com/showArticle.jhtml?articleID=17500289).

Part IV

Phase 2 Creator-Collaborator Capabilities

I have three goals for this section:

1. *To profile some exciting capabilities that many creator-collaborators will leverage as we move through the second phase of digital convergence.*
2. *To further your understanding of the tension between the creator-collaborator movement and the strategies and tactics of bottom-line–oriented corporate technology, media, and telecommunications providers.*
3. *To stir up your enthusiasm for the opportunities to profit personally from the individual and collaborative development and use of convergence technologies.*

In the next three chapters I'll offer a media-centric treatment of new capabilities, focusing in turn on text and image, audio, and video. Then, in chapter 17, I'll briefly summarize the technologies of mobile Internet. Finally, in chapter 18 I'll focus on media integration and interactivity, which is at the core of new capabilities that empower creators and collaborators.

These are fascinating times. Creator-collaborators now have considerable power, and I believe we will push technology in new directions that will ultimately enhance the lives of hundreds of millions of people throughout the world.

14. TEXT AND IMAGE: FIRST-PHASE STALWARTS

You could very often spot a phase 1 creator-collaborator by listening to them type. Rather than a peck here and a peck there, you would hear a flurry of keystrokes – the sound of a person "talking" through their keyboard. Indeed, text-based communication and information, together with the judicious use of images – largely as visual stimulus and aid to navigation – attracted millions to the Internet melting pot of ideas, culture, and commerce:

- The World Wide Web (which relied primarily on text and images) and text-based electronic mail were the killer applications that initially drove widespread global adoption of the Internet.
- Web images reflected the wide-ranging interests served by this new communications platform – from family photographs, to comic strips, to corporate "brochureware," to pornography.
- Web-based search engines, online chat, and online discussion forums utilized text to deliver useful new services and interactions.
- Distance learning applications were based primarily on text. For example, two leading Web course management systems, Blackboard and WebCT, utilized primarily text-based communication.
- Phase 1 e-commerce was based on textual and numeric information. For example, eBay (one of the greatest phase 1 creator-collaborator success stories with $24 billion in gross merchandise sales in 2004, up 60% over the $15 billion reported in

> 2002*) has depended almost exclusively on text and images. Many consumer-oriented e-commerce sites, such as Amazon.com and LandsEnd.com, have also relied exclusively on these two media types.

Text and images have proven to be a potent combination, and even as rich media becomes more integral to our electronic interactions in phase 2, text and image will continue to play a very important role. With that in mind, I offer a brief summary of a select set of text and image applications that are particularly interesting and potentially useful to many creator-collaborators: blogs, Instant Messaging (IM), and digital photography.

Blogs

> "These days, there are tech blogs and sex blogs and drug blogs and onanistic teenage blogs. But there are also news blogs and commentary blogs, sites packed with links and quips and ideas and arguments that only months ago were the near-monopoly of established news outlets."
>
> Andrew Sullivan, *Wired*, May 2002**

A blog (short for Weblog) is a continuously updated Web-based commentary or diary authored by an individual on a topic or set of topics, with reactions and comments from others who share the interests of the blogger. It is an interactive resource where a community of creator-collaborators can share information, ideas, and insights.

---

* According to a press release on the eBay Web site, *eBay Inc. Announces Fourth Quarter and Full Year 2003 Financial Results*, January 21, 2004 (http://investor.ebay.com/news/Q403/EBAY012104-712351.pdf).

** Andrew Sullivan, "The Blogging Revolution," *Wired*, May 2002 (http://www.wired.com/wired/archive/10.05/mustread.html?pg=2).

I think it is instructive to place the blogging phenomenon in the historical context of the Web and its relatively short lifespan.

Early on in phase 1, the Web provided a platform that creator-collaborators could use to create Web sites that were largely composed of text and images (appropriate for narrowband access). Many individuals, groups, and communities of interest quickly gravitated toward the Web for primarily text-based information sharing. The Web provided a platform for a new form of interactive self publishing.

As more and more people got up to speed on HTML, and as software packages like Netscape Composer and Microsoft FrontPage enabled relatively easy end-user creation and updating of Web pages, an explosion of Web content ensued. Eventually, many creator-collaborator self-publishers realized that maintaining a Web site, even with an easy-to-use WYSIWYG* Web editor, could be complicated and time-consuming.

Blogging software introduced in the late 1990s, such as Pitas (pitas.com) and Blogger (blogger.com), provided a much easier way to publish text-based information and commentary, in a format where reader comments could be easily accommodated as well. The simplest, easiest-to-use blogging software tools are interactive Web sites that enable the blogger to easily post new content while choosing among a number of predefined templates for layout, format, and color.

Blogging software enabled push-button publishing with an attractive look and feel and the opportunity for collaboration and interaction.**

Many creator-collaborators quickly recognized blogs as a useful platform for niche-oriented information sharing and interaction, creating an explosion of blogs, with topics ranging far and wide. The phenomenon has produced many high-quality resources, and many believe that the

* WYSIWYG: What You See is What You Get (pronunciation: "wizzy wig").

** A thoughtful explanation of the early history of the Weblog phenomenon is available at the *Rebecca's Pocket* blog, September 7, 2000 (http://www.rebeccablood.net/essays/weblog_history.html).

You can find a good description of a range of blogging software platforms at the Microcontent News Weblog, July 22, 2002 (http://www.microcontentnews.com/articles/blogware.htm).

general public will increasingly view blogs as a reliable source of information and insight.*

The blogging phenomenon is also interesting because of its timing. Dramatic growth in the number and use of blogs occurred during the demise of the dot.coms and the accompanying downturn in the tech sector. This was one further indication that the fundamental empowering factors of convergence – which enable human beings to interact, communicate, collaborate, and share information in new ways – were continuing to play out, regardless of the harsh lessons endured by venture capitalists, financial markets, and technology providers.

Instant Messaging

Instant Messaging is a peer-to-peer tool that enables instant, text-based communication between two Internet-connected individuals.

In 1988, Finnish Unix system administrator Jarkko Oikarinen developed Instant Relay Chat (IRC), a popular Unix-based utility that enabled synchronous text-based group discussion.** In an online "chat," you go to an open forum, called a "chat room" (also referred to as a "channel"), where the text you type, and the text typed by anyone else participating in the chat room, is transmitted for all to see. From within a chat room you can invite an individual into a private chat room where you can have a one-on-one chat.

* You can find a useful blog search engine at http://www.blogsearchengine.com. At http://www.weblogs.com you can see a list of recently updated blogs.

Conventional media providers have also recognized the growing importance of blogging, and many have cultivated their own weblog resources. These run the gamut from large media outlets like MSNBC, which now has several bloggers featured at http://weblogs.msnbc.com, down to local media outlets such as Syracuse.com, a companion to the Syracuse Newspapers, which features local bloggers at http://syracuse.com/weblogs/index.ssf.

** From the Internet Relay Chat (IRC) Web site (http://www.irc.org/history.html).

ICQ ("I seek you") software, developed in the mid-1990s, offered a similar synchronous text-based communication mode that spawned the widespread popularity of Instant Messaging (IM). While IM communications are similar to chat, there are a few notable differences:

- Instant messaging uses client/server interactions for identifying and initiating connections between users, but direct peer-to-peer interactions for communication between Instant Messaging clients.
- Instant messaging is designed to enable a private communication between two participants in its basic form, with the option of engaging in group conversations. This is the reverse of the chat room model, which offers group discussions, with optional private chat.
- IM has a mechanism that enables participants to keep track of regular IM partners who are online and available for messaging. In instant messaging lexicon this is referred to as a Buddy List. The industry term for this is *presence management*.

Other IM tools were developed on the heels of ICQ, including Yahoo! Messenger and MSN Messenger, but AOL Instant Messenger (AIM) became the clear leader in terms of overall number of users. Microsoft now has a new IM client, called Windows Messenger, which is bundled with the Windows XP operating system. America Online purchased Mirabilis, the company that had developed ICQ, for $287 million in 1998.*

One of the most interesting things about instant messaging is the incredible variation in the perception and usage of this tool across different age groups.

During the fall of 2001 I taught two classes. One was a class of mostly junior and senior undergraduate students. The other was a weekly three-hour class of adults enrolled in our MBA program in Corning, New York (home of Corning Inc.). One week I shifted schedules so that I

* Eric Luening, "AOL Acquires Instant Message Firm," *News.com*, June 8, 1998 (http://news.com.com/2100-1001-212004.html?legacy=cnet).

could have the two classes meet via videoconference. I asked the students to prepare interview questions to see what they could learn about the technology and its impact by probing the thoughtful perspective of a very different demographic.

Among the most interesting interview questions was one posed by one of the undergrads: "How many buddies do you have on your Buddy List?" Several in the Corning class had no idea what a Buddy List was, and most had never used IM. In the undergraduate group, all were IM users, and the Buddy List count ranged from dozens to well over a hundred.

While you may or may not be one who writes off instant messaging as a plaything of teenagers and twenty-somethings, you should begin to recognize that the next generation is indeed wired to this technology like millions of baby boomers (myself included) were wired for television.

Our young people have become accustomed to knowing who is online and who they can talk to at any given time. They've embraced presence management and now take it for granted. Once the Internet reaches out to cell phones, PDAs, set-top boxes, and other Internet devices and appliances, presence management promises to make it even easier to keep in close contact with social or business partners.

IM is not just a tool for social interaction – there is an increasing number of viable business applications for this emerging peer-to-peer technology. For example, some stock traders use instant messaging to keep in touch with their trading partners and clients; and many corporate IT support teams are turning to IM technology to provide improved service to troubled enterprise computer users. A May 2002 *InternetWeek* reader poll found that seventy-six percent of respondents (mostly corporate IT staff members) recognized the use of IM in their organizations, and twenty-six percent either supported or planned to support business use of IM software. Only twenty-three percent of respondents indicated that there was no use (many had explicitly banned IM for security reasons).*

* Richard Karpinski, "Water Cooler: Instant Messaging Security," *InternetWeek*, May 30, 2002 (http://www.internetweek.com/story/showArticle.jhtml?articleID=6406305).

The IM software originally developed by AOL, Microsoft, and Yahoo! was developed without standards-based protocols, which has effectively prevented communication between different IM clients. Unfortunately, market dynamics provided a disincentive to standardize IM communication and open up IM networks. For example, market leader AOL Time Warner has appeared to be very reluctant to open up its AIM network to other IM providers even though required by the Federal Communications Commission as a term of the merger of AOL with Time Warner (the FCC also restricted the introduction of new AIM capabilities).* As Microsoft's Windows Messenger and Yahoo! Messenger developed new capabilities not offered by AIM, and these clients began to gain some serious market share, they too began to shy away from the pursuit of open IM interoperability. Over the past few years, several significant yet largely ineffective efforts have also been undertaken to overcome the lack of IM interoperability.**

---

* Jim Hu, "AOL Not Sold on IM Interoperability," *News.Com*, July 23, 2002 (http://news.com.com/2100-1023-945909.html).

Bob Woods, "AIM and Interoperability: Not So Fast," *InternetNews.com*, July 24, 2002 (http://dc.internet.com/news/article.php/1432031).

As Windows Messenger and Yahoo! Messenger have now achieved significant market chare, the FCC has lifted restrictions on AIM. From Declan McCullagh and Jim Hu, "FCC Lifts AOL Messaging Limits, *News.Com*, August 20, 2003 (http://news.com.com/2100-1032-5065650.html?tag=nl).

** Trillian (http://ceruleanstudios.com) is an IM client that can communicate with AIM, Yahoo Messenger, and Windows Messenger clients, providing users with an interoperable option, although some features of other IM clients are not supported. Jabber (http://jabber.org) is a set of IM protocols designed to enable standardized instant messaging and presence management across the Internet. Unfortunately, as of this writing no mainstream IM client supports Jabber. SIP for Internet Messaging and Presence Leveraging Extensions (SIMPLE, http://www.ietf.org/html.charters/simple-charter.html) protocols have been developed by the IETF community to provide an interoperable underpinning for instant messaging. Odigo Messenger for Windows (http://www.odigo.org) enables communication with ICQ, AIM, and Windows Messenger users.

As of this writing it seems that this trend is finally beginning to reverse itself: AOL Time Warner and Microsoft have agreed to explore options for establishing interoperability between the two IM networks.* We can hope that increasing standards-based interoperability will emerge, enabling IM to take center stage as a universal, mainstream communications platform. Instant messaging technology has begun to incorporate support for a wide variety of features including support for audio and video. Combine these emerging rich media capabilities with presence management, the incredible support for IM among younger global citizens, and mobile access, and you may have a formula that could propel IM into a phase 2 killer application. It has the potential to emerge as the cornerstone of a new mode of rich human interaction via an increasing variety of electronic devices, from ultra-portable handhelds to desktop PCs.

Digital Photography

Next time you're in an electronics superstore, saunter up to the digital photography counter and hang out for a minute. What will you find? A gaggle of tattooed, sagged, and pierced millennials ogling the cameras? Not likely. You'll probably encounter a diverse group of individuals who all share a love of photography, the time-honored medium that binds us together while providing a visual history of our time on earth. While digital photography has quickly evolved to offer great new imaging capabilities, it is also fascinating precisely because – unlike many other emerging convergence technologies, such as instant messaging – this is an application realm that is on the radar of a very broad cross-section of humanity.

* Joanna Glasner, "Getting the Message About IM," *Wired News*, June 2, 2003 (http://www.wired.com/news/business/0,1367,59046,00.html).

The Golden Rule of Technology at Work

In the first book I introduced the golden rule of technology,* which is clearly at work in the realm of digital photography.

At the Whitman School we bought our first digital camera in the mid-1990s. It was a Kodak that took only reasonable quality pictures at a relatively low resolution of 640 x 480 pixels.** It cost close to $500. You downloaded pictures through a PC serial port, which was slow. The software was clunky, and images were in a proprietary format that could only be read by the Kodak software (you had to export photos before opening them in an image editor such as Photoshop).

In late 2001 I bought another camera for my own use. I ended up buying a Fuji FinePix 2800 for about $400. It has a 6x optical zoom and a maximum resolution of 1600 x 1200 pixels (roughly 2 megapixels).*** The camera has automatic, fill, or redeye-reduction flash options, and it captures images in compressed JPEG format with fine, normal, and basic quality options. Images are saved on removable Smartcards, and the software allows me to easily and quickly download the standard JPEG images via a USB port. It came with an 8-MB Smartcard which enabled me to store somewhere around twenty high-resolution compressed JPEG images. I went to a discount store and bought a 64 MB card for around $75, so I can now shoot over one hundred high-resolution images between downloads. I'm thinking of purchasing a Smartcard reader which I could attach to my PC for quick, easy transfer of images without draining camera batteries.

Professional photographers have also seen the golden rule at work. When I was writing the first book in the late 1990s, I had a casual con-

* The power, functionality, and ease of use of any digital technology will continue to increase as the cost continues to decrease.

** Pixel is short for picture element; as explained in the first book, a digital image is a matrix of pixels, where each pixel is assigned a color value.

*** As of February 2004, Fuji is selling a newer FinePix model that has 3 megapixels and a 10x optical zoom for under $300.

versation with Steve Sartori, Syracuse University's staff photographer, to get his thoughts on digital photography. At that time Steve was excited by some potent benefits – in particular the instant access to images with no messy developing, and the power of digital editing tools. But like other professional photographers I've talked to, Steve had recognized the limitations of digital cameras and tools such as inferior quality (at reasonable price points) and the difficulty of capturing good action shots.

The passage of a few years has seen significant improvements to digital cameras and related technologies, and Steve now finds himself migrating much of his work to digital. Recently, we retained his services to take pictures of our entire freshman class for a photo database, and sure enough he used a digital camera. Steve also helped us set up Adobe Photoshop scripts to automatically convert the hundreds of freshmen photographs to a format suitable for access by our faculty through a Web interface. I spoke to him about his use of digital cameras, and he told me that he hadn't shot anything but digital for many months.

Mary Peery, who was vice president and general manager of HP's digital imaging group when I interviewed her in the spring of 2002, told me that there is no doubt that the transformation to digital is in full gear. She cited annual industry-wide revenues of $15 to $20 billion, and she forecast a tripling of those figures in the next few years. Clearly, the writing is on the wall: Conventional film photography used to be the norm, with a few select professional applications and limited use by consumer enthusiasts; but the balance is now shifting, and we will soon find that digital photography is the norm, with film-based photography used for a few specialized applications.

Perfect for Many Creator-Collaborators

When you get that first digital camera, and you begin to take pictures and then explore the possibilities for transforming and sharing your digital images, it's tough not to get excited. This often leads to heightened expectations for the ways in which technology can enhance life and career experiences. Not surprisingly, digital photography serves as a launching pad for many creator-collaborators.

If you have an interest in exploring new media tools and technologies, digital photography is something you should consider, if you haven't already. With a bit more effort than what a passive consumer might muster, you can experience a low-cost tool that empowers you with high-quality results. I guarantee that with a taste of what you can do with digital photography, you'll quickly warm to the potential of this and other convergence technologies.

If you work in a large organization you'll probably find that digital photography promises compelling individual and workgroup applications in several business contexts. This technology is typically outside the realm of central IT controls, and a straightforward business case can put the technology directly in the hands of you and your co-workers. I encourage you to think creatively and go for it!

Here are a few insights that may help you if you do elect to pursue this ready-for-prime-time technology.

1. Print Quality

Most people have seen inkjet printouts of digital photographs that just plain look bad, and look even worse once they've been sitting around for any significant amount of time. But here again, the technology has improved to the point of offering a solid solution. If you have a fairly high-resolution camera, buy a decent color inkjet printer and some good photographic inkjet paper. You can generate really nice prints. Spring for a photo printer (which, like everything else digital, continue to decline in price) and you can print out 5x7s or 8x10s that can rival what you get from a conventional photo processing center.

2. The Resolution Conundrum

If you buy a good consumer-grade digital camera and a decent printer and paper you will find that you can generate good-quality prints. But you will also find that working with high-resolution images on a PC

– displaying them in full resolution, posting them on a Web site, or sending them via email – can quickly become painful.

To get quality hardcopy prints from a digital image, you must have a high-resolution image. This means getting as many pixels in the image field as is possible. Today's garden-variety digital cameras come in 3 and 4 megapixel versions (1 megapixel is approximately one million pixels), enabling good-quality prints at relatively large print dimensions. It follows, then, that people who buy these cameras for personal use seek to get the most pixels for the buck.

However, when displaying digital images on a computer monitor you encounter some serious constraints. There is a limited matrix of display pixels on a computer's monitor; and high-resolution images won't fit on a screen configured at 800x600 or 1024x768, which are common resolutions for Windows computer users. You end up scrolling around just to see the entire image. Higher resolution images also require more bits, and this can become problematic when it comes to storage and network transmission.

If you go out and buy a decent camera, you will need a strategy for capturing images for printing as well as for computer display and Internet sharing. One strategy is to shoot images that you want to print at high resolution; and shoot those images you will share on the Internet at a lower resolution. While this is a consumer-friendly strategy because it is easy to execute, it is problematic if you later decide you want to print out your low-resolution images or do any significant cropping.

An alternative strategy is to always shoot high-resolution images. Once you download them you can print them out with good results. You can also peruse the thumbnails looking for those images you may want to share on the Web or via email. Then, with a decent image viewer or editor you can easily edit, crop, resize, and compress the images so they are suitable for viewing on a computer monitor and transmitting via the 'net. This is the strategy that I use. And while this is a bit more work than the consumer-friendly strategy outlined above, for many creator-collaborators it's not a big deal given the powerful yet fairly easy-to-use software that is now available.

3. Powerful Software

One of the huge differences between conventional photography and digital photography is the power to modify and enhance images after you've captured them. The power, flexibility, and functionality of inexpensive and relatively easy-to-use image viewers and editors is downright amazing. Here is a quick profile of a few packages you may find useful.

- *Picasa* is a very modestly priced image editor and slide show generator available at www.lifescapeinc.com/picasa. With Picasa, you can easily remove redeye or optimize the colors and contrast in your images. This package also has a great cropping tool that allows you to crop and view your images without actually modifying the original picture. A retired professor in the Whitman School was given a digital camera as a retirement present, and since then he has traveled the world and documented his extraordinary travels with pictures. He is far from a computer jock, but he has had no trouble using Picasa to organize and display his cropped, optimized images – and the results are very impressive!
- *IrfanView* is an image viewer you can download at www.irfanview.com. With IrfanView you can take a set of slides in a folder or set of folders and convert them in a single batch operation. This is useful for resizing so you can shoot at high resolution and convert your photos in one operation to a lower resolution suitable for Web or email transmission. IrfanView also has some basic editing tools and a great slide show mode. I've used the slide show many times for quick-turnaround slide shows (e.g., at the reception following the ground-breaking for our new Whitman School of Management building).
- *Photoshop*, from Adobe, Inc. (www.adobe.com/photoshop) is the gold standard in image editing. With an extensive set of tools

> for manipulating images, Photoshop is used by all types of professional photographers, graphic artists, media developers, Web site developers, and anyone else who is serious about the effective and professional use of digital imagery. While Photoshop is pricey, it offers incredible control and creativity. The complexity and price of Photoshop can be a bit overwhelming for creator-collaborators, so Adobe has taken a subset of Photoshop capabilities and features and bundled them with some user-friendly tools in a less expensive version called Photoshop Elements. Photoshop Elements includes the most commonly used features of Photoshop – image layers, advanced selection tools, filters, and so on – along with some user-friendly add-ons like a wizard for automatically exporting photos to a Web site.

With Photoshop or Photoshop Elements, I can easily make minor adjustments to my images, or get a little creative as suits my inclination and available time. Below is a "before" and "after" image that I use in a Photoshop demonstration in some of my lectures and seminars to show people the power and ease-of-use of this and other digital editing tools (I usually do this as part of a demo to create a digital video where I also use SoundForge, an audio editor, and Adobe Premier, a prosumer-grade video editing package).

I start with an image created by taking a quick digital snapshot of Bob Ranger (a School of Management computer support person) in front of an office window that looks out over the Syracuse University campus. You can see Crouse College, one of the landmarks of the Syracuse University campus, in the background.

Before

After

I perform the following functions, which are really easy to do:

- Crop the image to remove unnecessary information, and resize it to reduce storage (and Internet transmission) requirements
- Select the sky and fill with an orange hue at 40% opacity to create an "Orange" sunset over Syracuse
- Use the clone tool (a.k.a. the rubber stamp tool) to remove the camera flash reflection that appears over Bob's shoulder
- Add a royal blue text title, and give it a very slight drop shadow
- Save it in a compressed JPEG format for Web or email use

I'm no Photoshop expert, but I've learned enough to be dangerous!

15. AUDIO: LEADING THE CHARGE

The evolution and integration of digital audio* technology in a broad range of hardware and software has enabled audio to be one of the first real-time rich media types to shine in the early days of phase 2.

In the last two years I've twice taught "Digital Convergence and Digital Commerce," a course in our iMBA program (our popular distance-learning MBA program). Both times I presented material to students around the globe using an on-demand streaming audio narration of PowerPoint slide shows. I also had student teams create on-demand presentations consisting of streaming audio and PowerPoint presentations. While video might have been better, audio is eminently more doable. Compressed audio streams can be effectively transmitted over even narrowband Internet connections. Each student had a PC with a sound card and some kind of Internet connection, and at least one student on each team either had or acquired an inexpensive microphone. Software programs for encoding the audio streams are free downloads from Real Networks and Microsoft (for this course my students and I used Real Networks' Helix Producer to encode audio). The narrated slide shows used in this course have worked reliably with far-flung students in Greece, Japan, Bermuda, British Columbia, Texas, and Maine.

This is but one example of a converged audio application that just works. I'll bet you too have encountered some sort of compelling Internet or Web technology that effectively incorporates sound. As we move toward a time when rich media is more integral to our Internet interac-

* You may recall from the first book that sound is compression/decompression waveform. The human voice, musical instruments, and the tree that falls in the woods all create sound waves that have varying characteristics in terms of frequency, which determines pitch, and amplitude, which determines volume. Given the relative maturity of digital compression technology, digital representation of sound is now possible with a relatively modest amount of digital bits.

tions, we find that audio technologies and applications are leading the charge.

In this chapter I'll briefly profile two of the most important audio-centric application realms: Internet telephony and music on the 'net. The early phase 2 evolution of these two realms is both interesting and instructive.

Internet Telephony

A conventional phone call is a synchronous, point-to-point voice conversation. The use of the Internet as a medium for point-to-point, synchronous voice conversations is referred to as Internet telephony or Voice over IP (VoIP). This first became popular in the mid-1990s when standard desktop PCs (with sound cards), Internet access, and a downloadable software program like Net2Phone or the Web-based Dialpad enabled free low-quality voice conversations across the 'net.

Internet telephony has come a long way since then:

- Voice quality has improved.
- Walkie-talkie–style communication, referred to as half-duplex, has been superseded by full-duplex communication – so that you and the party you're calling can both speak and be heard at the same time.
- Protocols, such as the IETF Session Initiation Protocol (SIP) and the International Telecommunications Union's H.323 suite of voice over packet network protocols, have been developed to enable functional, interoperable Internet telephony services and applications.
- Much-needed bandwidth has become more readily available in the backbone, the enterprise, and the last mile – enabling clearer, more reliable 'net phone connections.

- Internet phone handsets can provide portability throughout a corporate LAN or home network.
- Internet telephony providers have created gateways to the public-switched phone network (PSTN), so you can now place a call through the Internet to almost any conventional phone number.

While many creator-collaborators used first-generation Internet telephony technology, it was just too cumbersome and unreliable for most consumers, and many telecommunications carriers saw this as a real threat. A group of long distance companies (America's Carriers Telecommunication Association or ACTA) quickly pursued a regulatory challenge by petitioning the FCC to regulate Internet telephony as a common carrier telecommunications service.* Since those early days the use of public and private internet networks for voice conversations has caught on not just among creator-collaborators, but among consumers, corporations, and telecommunications providers as well.

Here are some examples:

- Verizon has begun work to convert its circuit-switched network to Voice over IP, while Qwest has already begun to offer VoIP-based long distance services and AT&T has plans to use VoIP to enter local calling markets.**
- Net2Phone (net2phone.com), one of the early providers of PC-to-PC phone calls across the public Internet, has evolved to provide a range of technologies and services, including low-cost long distance services, network service for voice over cable tele-

* Mark Naftel, "The Promise and the Threat of Internet Telephony and Regulation," The Phoenix Center Web site, reprinted from *The Computer Law and Security Report*, November/December 1996, Volume 12, Issue 6 (http://www.phoenix-center.org/library/telephony.html).

** Ben Charny, "Verizon Begins Last Leg of VoIP Effort," *News.Com*, January 7, 2004 (http://news.com.com/2100-7352-5136928.html).

vision networks, and adapters for plug-and-play use of conventional phones over IP networks.

- USA Datanet is a Syracuse-based company that implemented a high-speed TCP/IP network to provide low-cost long distance service in the northeast ($.99 for calls in the northeast). The company is expanding its "carrier-class" IP voice network – which utilizes technology from Sonus Networks, a leading provider of Voice over IP network infrastructure technology – and it will soon provide local calling service to complement its long distance service.*

- Vonage (vonage.com) is a consumer-oriented broadband Internet telephony provider that offers a slew of features such as voicemail with email notification of new messages, virtual phone numbers where you can pick your area code, 911 emergency service, call forwarding, and simultaneous calling to your Vonage number and one other U.S. phone number.

- Skype (skype.com) is a PC-based peer-to-peer Internet voice application developed by the creators of the popular Kazaa P2P file sharing program. It enables free encrypted phone calls to other Skype users and allows users to place sixteen calls on hold and to carry on conference calls with up to five other users.

- Large network and telecom equipment providers such as NEC and Cisco now offer reliable VoIP technology that small and large organizations are implementing over local and enterprise networks – enabling them to scrap their conventional voice technology and move to a single converged packet network for voice and data.

* Tim Knauss, "USA Datanet Takes on the Big Phone Companies," *The Post Standard* (a Syracuse Newspapers publication), February 8, 2004.

- The Free World Dialup project, founded by Jeff Pulver (pulver.com), provides free PC-to-PC phone service over the public Internet. While there are relatively few users of this Internet-only voice service as of this writing,* the FCC has decided that this service will not be required to pay the same access fees required of conventional phone carriers – paving the way for broader deployment and adoption of free phone services.**

Corporate and Telco Cost Savings

Among many corporations and telecommunications carriers, the VoIP end-game is really simple cost reduction. Large corporations that have geographically dispersed offices can move Internet phone calls over the corporate intranet to avoid costly long distance calls. Telecommunications providers can cut costs by eliminating expensive circuit switching equipment and moving to more efficient delivery of voice via IP-based packet networks. Businesses building new facilities can provide phone and data service using a single physical network cable plant. Corporate implementations can enable phone handset adds and moves that are simpler and less costly than on a conventional enterprise phone network.

Major communications and technology providers from all sectors (e.g., Qwest, Cisco, Nortel, Lucent, Lucent spinoff Avaya, Time Warner, and AT&T) now recognize the advantages of Voice over IP technology and are working to find the products and services that will return handsome profits as conventional telephony cedes to this converged alternative.

---

* As of February 13, 2004, the Free World Dialup service had over 156,000 users from over 185 countries. From the Free World Dialup Web site (http://www.fwd.pulver.com).

** Stephen Labaton, "F.C.C. Begins Rewriting Rules on Delivery of the Internet," *New York Times*, February 13, 2002 (http://www.nytimes.com/2004/02/13/technology/13net.html).

While the potential for cost savings is a dominant driver of Internet telephony, there are also some useful and compelling features and applications which may take hold once widespread migration to Voice over IP is further along. A few examples:

- Phones with small screens – that integrate voice with additional 'net-based services and information.
- Unified messaging – which provides easy access to IP-based phone messages along with messages from other messaging applications, such as fax and email.
- Portability – log on to any phone handset on the enterprise network to take your calls; or use your notebook PC as a softphone anywhere on the planet where you can plug it into the public 'net.
- Web-based click-to-talk services – that enable Web surfers to initiate voice calls to remote sales agents or customer support representatives (who can then take control of the Web interface for multimedia demos or training sessions).

Still a Little Disappointing

With the promise of current Internet telephony comes the reality of implementing a relatively immature technology. Some significant challenges relate to network traffic management, Quality of Service, and reliability. These issues are being addressed, but as with most new technologies, this occurs in a series of baby steps rather than a smooth, fast transition.

When I first started to write this book, Syracuse University's contract for on-campus phone service was soon to expire, and the university evaluated Voice over IP as an alternative. The central IT group had some significant concerns including the considerable cost of upgrading the entire campus network to ensure the QoS that VoIP requires, and concerns over reliance on an open and shared network where hackers occasionally wreak havoc. The university decided to re-sign with its telephony system provider, and re-evaluate when the contract next comes up for renewal in a few more years.

When considering NEC VoIP technology for our new building, I interviewed an IT manager at a midwestern metropolitan state university that had implemented this technology. She told me that while they liked some of the new services the system provided, they also had experienced some difficult-to-diagnose and -solve problems (such as sporadic static and echo) and they had to spend a considerable amount of time responding to user complaints.

While the trade press has been fussing over VoIP for some time now, this creator-collaborator still finds Internet telephony technology a bit disappointing overall. Successful enterprise deployments are growing in number, but they are still fairly few and far between. For you and me, VoIP offers a means of saving money on long distance calls but little else – and long distance price wars have already brought down the cost of conventional and wireless long distance calling to historically low levels.

Hopefully, as VoIP technology gets integrated into a wider variety of communication tools – such as Instant Messaging software and Web presentation platforms – it will create compelling and useful capabilities built on the integration of real-time voice communication with video- and text-based collaborative tools and interfaces.

The Second-Class 'Net

In the first book I indicated my hope that Internet telephony would stimulate major improvements to the public Internet, especially in the ability of this now-pervasive communications platform to reliably support real-time transmissions and QoS. But Internet telephony has been largely pursued in the context of service provider and enterprise networks, not in the context of the open, public Internet. Unfortunately, as we discussed in chapter 13, end-to-end improvements across the public Internet infrastructure appear difficult at best. There has been significant progress, so I have seen less evidence of ugly congestion and delay in the recent past, but these issues have not yet been eliminated.

The nomenclature of packet-based voice services has mirrored an increasing emphasis on the deployment of advanced communications services over proprietary IP networks rather than the public Internet. The original Internet-based voice capabilities – PC-to-PC voice calls across the Internet – and the technologies and services that followed on the heels of those early connections, were most often referred to as "Internet telephony." Now that term is almost never used, since so many of the technologies and services have been targeted for enterprise intranets and other nonpublic IP networks. Instead, "Voice over IP," or VoIP, is the most often used term for the latest digital voice technologies and services. And more recently, packet-based voice companies and telecom industry leaders (such as Reed Hundt, former head of the FCC[*]) have begun to refer to this packet-based voice service as Broadband Telephony or Broadband Voice (BBV) – further removing this technological phenomenon from the sphere of the public Internet.

My hope is that Internet telephony applications such as Skype, Vonage, and Jeff Pulver's Free World Dialup project, along with other similar grassroots public Internet technology movements, such as Internet radio (described below) and multimedia messaging, will begin to experience significant success and enter the mainstream. This would help to reinvigorate the development of new interactive rich-media services that are designed from the ground up to be open and freely available on an improved public Internet.

[*] Paul Kaputska, "Why VoIP Needs a New Acronym," *Networking Pipeline*, January 26, 2004 (http://www.networkingpipeline.com/showArticle.jhtml?articleID=17500934).

The Digital Music Revolution

Digital convergence has altered the way that human beings can generate, acquire, experience, and share media, and music is the first rich media entertainment content realm to be forever changed by new expectations for when, where, and how we can enjoy it.

While many passive consumers have yet to benefit from new converged music options, enhanced capabilities are routinely used by many creator-collaborators. Beyond that, many recent developments within the new digital music movement have been driven by creator-collaborators who have created innovative new technologies, resources, and mechanisms for creating and sharing music. Creator-collaborators are a powerful force, and a quick review of the digital music movement provides a useful and instructive example of creators-collaborators in action. Before we begin to explore creator-collaborator efforts in a little more detail, let's first reflect on the overall impact of convergence on popular music.

Popular Music and the Star System

Most people love music. It evokes emotion. It inspires. Popular music defines each generation.

For decades, the creation, marketing, distribution, and sale of popular music has been based on what I call the star system. This is a system where well-organized, well-funded, and heavily promoted efforts to create popular music stars have been successful in driving millions of purchases of albums and CDs – from a relatively small set of artists.

On one hand, this makes sense. Human nature is to gravitate to shared experiences and things that are popular. Repetitive exposure through well-coordinated conventional media outlets results in familiarity and, often, increased enjoyment of a particular song.

On the other hand, the star system uses radio, television, and other mass media to drive our attention to a miniscule portion of available artists and recorded music. Listen to almost any popular music radio station and you will encounter an often annoying repetition, called a playlist. Turn on MTV, BET, CMT, or VH1 and you'll find a similar approach to featured artists (unfortunately, some of these networks have dramatically decreased airtime for music videos). While repetition creates familiarity, which is sometimes desirable, it all too often causes boredom and frustration as well.

The star system has generated significant revenue from consumer purchases of physical media (vinyl LPs, then 8-track and cassette tapes, and now CDs) to acquire music for personal use. I once heard a music industry insider describe this as a "bait and switch" strategy: the industry promotes singles on radio and television, but then sells mostly CD compilations which feature at least a dozen songs. How often have you bought a popular album or CD because you had heard and liked just one or two songs?

A New Natural Order

The Internet, the Web, peer-to-peer file sharing (popularized by Napster), and a wide range of other convergence technologies have ushered in a new natural order for popular music – i.e., a new set of factors that change the way music can be shared and enjoyed.

- The use of physical media to disseminate music recordings is no longer necessary. We can easily copy and share digital music. And we now have available to us an incredibly diverse and voluminous set of recordings that are in digital format and 'net accessible.
- I pointed out in Part I that convergence technologies have enabled what the Cluetrain Manifesto refers to as a market conversation. Applying this to popular music, our tastes can be informed not only through television, radio, and other mass media, but through electronic interactions with other music lovers through fan sites, music communities, instant messaging, peer-

to-peer sharing tools, and other forms of electronic "word of mouth." This challenges the control of conventional mass media and mass marketing.

Flexibility, choice, and market conversations are increasingly important drivers of the digital music revolution. We can hope that the music industry will quickly realign its model for promoting, distributing, and selling music recordings accordingly – but I'm not counting on it. The music industry will likely continue to focus its energies on a fight against these new fundamental forces; and the digital music revolution will continue to be a creator-collaborator–led phenomenon where the new natural order enables a richer and more personalized and flexible mode of music enjoyment.

Let's focus in a bit on the forces that have created this new natural order and review some of the outcomes that have already resulted.

Musical Creator-Collaborators

The golden rule of technology clearly applies to digital music technologies. Cheaper, better, and more capable hardware and software enable professional and novice musicians alike to collaborate and create quality digital music compositions without requiring a gargantuan investment of human and fiscal resources. The results can be easily shared via CDs or Internet transmission. This offers new creative outlets while leveling the playing field (at least to some degree) for tech-savvy independent musicians working hard to promote and distribute their music to a growing audience.

Here are some examples:

- Many new music artists are using notebook and desktop computers as primary platforms for the creation of new music. With a software-based synthesizer, a boatload of digital music samples taken from prior works, and some advanced multitrack music software, tech-savvy music artists can work anywhere, anytime while churning out some very exciting new music. Brian Transeau, better known as trance music pioneer BT, is one such artist. Like many other music artists in the age of convergence, BT works extensively with digital music. He can create and edit his work on his laptop while on the go, as was the case with two to three thousand vocal edits which BT made – during an airplane flight – to "Pop," the first track on *Nsync's *Celebrity* album.[*]

- Nick Drozdoff is a Chicago-based jazz trumpeter and recording artist (and high school physics teacher) who has created dozens of high-quality recordings using a variety of musical instruments and some desktop computer hardware and software – all in his basement studio. He posts his music at the Ampcast.com community, where listeners can stream or download individual songs.[**] Electronically-empowered network-accessible artists such as Nick Drozdoff offer a wide range of musical styles to suit what I hope will be an increasingly diverse taste for music.

- Right after I wrote the first book, I was checking out the Web site of Public Enemy, the rap group whose front man Chuck D. has been a vocal supporter of digital technologies that enable many artists to interact more directly with their audiences.[***] I posted a message about digital convergence and digital music on the "Enemy Board," a discussion forum on the Public Enemy Web site. I was subsequently contacted by a group of rappers who collaborate across the 'net to make their own rap music.

---

[*] Mark McClusky, "The Incredible Shrinking Studio," *Wired News*, October 2, 2003 (http://www.wired.com/news/digiwood/0,1412,60639,00.html).

As of February 2004, BT's *Emotional Technology* CD could be streamed in its entirety at http://www.briantranseau.com.

[**] Nick's site can be found at http://www.ampcast.com/music/23689/artist.php.

[***] David Kushner, "Chuck D MP3 Won't Kill Labels," *Wired News*, February 24, 1999 (http://www.wired.com/news/culture/0,1284,18103,00.html).

> One person lays down a beat, then sends that digital recording on to another member of this virtual music group who will lay down some vocals, and she then passes it along to another member of the group who adds another musical element, and so on – until the song is complete. They sent me a few cool tracks that I linked on my Web site.

Music community sites like Sonic Garden (sonicgarden.com), Ampcast, and HipTingle (hiptingle.spydigital.com) promote independent artists, provide access to new music, and build community not only around mainstream genres but around some very interesting niche genres that offer new listening experiences for many of us. I suppose we'll always have a top forty, but it's become very clear to me that the top forty artists aren't the only gifted artists out there.

For the many among us who aren't all that musically inclined, the power of digital music lies largely in our ability to easily access and share the music we enjoy.

Accessing and Sharing Digital Music

MP3* compression technology enables conversion of CD-quality digital music files to compressed versions that retain much of the original quality, in terms of sound fidelity, but are many times smaller, in terms of file size. You can take a conventional music CD, put it in your PC and use a software program, such as Winamp Pro, to quickly and easily convert the original digital recordings on the CD to MP3 files on your hard drive (this is called "ripping"). While a typical CD sound recording can

* I described media compression in the first book. MP3 was originally developed as part of the Motion Picture Experts Group MPEG-1 standard, which was developed for compressing motion video with accompanying audio. The term MP3 actually refers to the Layer 3 audio specification of the MPEG-1 compression standard.

easily be dozens of megabytes, a typical MP3 song is just two or three megabytes and can be downloaded through a broadband connection in just over a minute (something on the order of ten minutes to download through a dialup narrowband modem connection).

Among young people, it seems that music sharing via the Internet has become a killer application for broadband. An undergraduate course I taught in the spring of 2002 had forty-eight juniors and seniors. Most lived in apartments – and all but one student had a broadband connection. Sharing MP3 music files was the number one use of the broadband connection.

In the Beginning, There Were Creator-Collaborators

The first digital music sharing Web sites were launched by entrepreneurial creator-collaborators, such as Rob Lord and Jeff Patterson, who founded the Internet Underground Music Archive* (which was founded in 1993 and acquired by the digital music company that became EMusic in 1999). Another entrepreneurial creator-collaborator, Michael Robertson, subsequently launched MP3.com,** which became the largest music file sharing site before it was hit with copyright infringement lawsuits and was eventually sold to Vivendi Universal. At these sites, artists posted their songs to get some exposure and increase their fan base; and at some sites music loving creator-collaborators posted popular songs ripped from copyrighted CDs.

* For more on how these two University of California at Santa Cruz students launched the first digital music sharing Web site, see the University of Santa Cruz *Review Magazine*, "'Net Rebels Shake Up Music Industry," Winter 1995 (http://www.ucsc.edu/news_events/review/text_only/Winter-95/Win_95-Net_rebels_shake.html).

** For more on how Michael Robertson started MP3.com see San Diego State history Professor Steve Schoenherr's Recording Technology History Web site (http://history.acusd.edu/gen/recording/robertson.html).

Creator-collaborator Shawn Fanning was a student at Northeastern University during this early phase of the MP3 music sharing phenomenon. He liked sharing MP3s via Web sites like MP3.com, but he found these sites cumbersome and inefficient for finding and downloading all the music he wanted to hear. He decided that a peer-to-peer service (peer-to-peer is described in more detail in chapter 11) might work better, since it would eliminate the constrained choice and download bottlenecks that were all too common with typical Web client/server solutions. He programmed Napster as a peer-to-peer (P2P) music sharing tool and released it to other MP3 enthusiasts. It was an instant hit – millions downloaded the software within months, and before long almost any popular song you could think of was available as a free Napster download.

Today, with a desktop PC and a library of MP3 recordings, which can be ripped from CDs or downloaded using P2P software and an Internet connection, you can:

- Create playlists that enable you to play your favorite songs in the order you choose.
- Copy or "burn" your playlist onto a conventional CD (provided your PC has a CDR or CDRW drive) so they can be played in any standard CD player.
- Burn several dozen of your favorite MP3s onto a data CD, which are playable in some CD players and most DVD players.
- Load a sizable music library onto a portable MP3 player that is the size of a deck of cards.
- Share MP3 songs with other music lovers across the Internet.

While the flexible music enjoyment that this enables is wonderful in many ways, it has stimulated widespread sharing of copyrighted material.

The Industry Fights Back

The star system of heavily promoted popular music had been good to those very few artists who made it big, and the illegal sharing of copyrighted MP3s threatened to hurt these artists where it hurts the most: in the wallet. Some popular music stars, such as heavy metal band Metallica, fought the open distribution of MP3 recordings. Other artists, such as Courtney Love, of Hole, came out in favor of the more open and direct connection between artists and fans that Internet-based MP3 file sharing could enable. She argued in a speech at the Digital Hollywood online entertainment conference on May 16, 2000,* that only a very small percentage of the proceeds from CD sales actually ended up in the artist's pocket anyway.

The Recording Industry Association of America (RIAA) filed lawsuits against Napster and MP3.com to cease sharing of copyrighted music and hopefully end the threat to music industry business models. When these suits effectively disabled MP3.com and Napster as prime sources for free popular music content, new peer-to-peer music sharing tools quickly sprouted: Gnutella, LimeWire, BearShare, Morpheus, WinMX, Kazaa, and Xolox enable peer-to-peer entertainment file sharing with some useful new functions. Most of these new tools lack a central content directory, which was a focus of the legal action that shut down Napster; and some of the companies that provide these new tools are based in countries where there is no copyright law or where copyright violations are a low priority. While record companies will continue to file lawsuits to shut down file sharing services, success is far from assured.

Lawsuits are not the only tactic being pursued by record companies. The record companies have also launched new download services which offer fee-based alternatives to illegal file sharing (Figure 1).

---

* The complete text of this speech is online: "Courtney Does the Math," *Salon*, June 14, 2000 (http://dir.salon.com/tech/feature/2000/06/14/love/index.html or http://archive.salon.com/tech/feature/2000/06/14/love/print.html)

Figure 1. Internet Music Services*

Real Rhapsody (Real Networks)	RealAudio
iTunes Music Store (Apple)	MPEG-4 AAC
MusicMatch Downloads	Windows Media
Napster 2.0 (Roxio)	Windows Media

These services promise higher-quality music recordings than the creator-collaborator–ripped MP3 recordings available on peer-to-peer sharing services (if you've downloaded many MP3s from Napster or any other peer-to-peer service, you've likely encountered inaccurate names, songs with playback anomalies, and peer-to-peer virus transmissions). However, these services also implement usage policies that may limit what you can do with recordings once you've downloaded them. For example, they may limit your ability to burn copies or to share your music with other users. While the largest services have hundreds of thousands of tracks available, they still lack the broad range of music available on P2P networks. The variation of format and software player requirements across providers can also complicate usage.

The music industry has also begun to engage in something called *spoofing*, which is purposely placing bogus copies of popular songs, called *decoys*, on peer-to-peer networks. This is an effort to further degrade the peer-to-peer experience and drive people to the new industry-supported music services.**

When lawsuits aimed at P2P software and service providers and all of the other efforts, including those described above, failed to kill the

* For more information on these services see the *PC Magazine* summaries of these services (http://www.pcmag.com/category2/0,4148,25054,00.asp) and Tony Dreier, "Online Music Stores: Music To Your Ears?" *PC Magazine*, October 29, 2003 (http://www.pcmag.com/article2/0,4149,1366688,00.asp).

** Dawn C. Chmielewski, "Music Industry Swamps Swap Networks With Phony Files," *San Jose Mercury News*, June 27, 2002 (http://www.siliconvalley.com/mld/siliconvalley/3560365.htm).

sharing of copyrighted music recordings, the Recording Industry Association of America decided to go after individual violators. In 2003 hundreds of lawsuits were filed, and some reports indicate that illegal sharing of copyrighted music has finally begun to subside. A study from comScore and Pew Internet Life found that the number of Americans downloading music files on the Internet has dropped by half, and that the number of people downloading on any given day since the lawsuits began has also plunged.*

Many digital music enthusiasts continue to use peer-to-peer software such as Kazaa to download and share popular music. They have just gone underground, according to industry experts, who maintain that file swappers are more secretive and may share less frequently – but music file sharing maintains a pace that is still many, many times the pace of traffic seen on legal download sites.**

Ironically, the top Yahoo! Search term in 2003 was Kazaa – while several mainstream media properties filled in most of the other top ten spots (Figure 2).

Figure 2. Yahoo! Top Searches of 2003***

1. Kazaa	6. Eminem
2. Harry Potter	7. WWE
3. American Idol	8. Paris Hilton
4. Britney Spears	9. NASCAR
5. 50 Cent	10. Christina Aguilera

* Lee Rainie, Mary Madden, Dan Hess, Graham Mudd, "The Impact of Recording Industry Suits Against File Swappers," January 2004 (http://www.pewinternet.org/reports/pdfs/PIP_File_Swapping_Memo_0104.pdf).

** Associated Press Report, "Song Trading Still Popular Despite Suits," *New York Times*, February 22, 2004 (http://www.nytimes.com/aponline/technology/AP-Downloading-Music.html).

*** From the Yahoo! Web site (http://search.yahoo.com/top2003).

Internet Radio

Conventional radio stations embraced the Internet relatively early on, with many streaming live programming on the 'net. Radio execs reckoned that their broadcast rights for local radio-wave transmission also applied to Internet streaming, and while there were some radio programs that were explicitly precluded from Internet transmission, such as Rush Limbaugh's daily talk show, many popular music stations were streaming out onto the 'net in the U.S. during the late 1990s.

This enabled us to:

- listen to a favorite local station in the car on the way to work, and then tune in on a desktop PC upon reaching our desks.
- tune into a childhood hometown radio station
- catch the radio broadcast of our alma mater's big game
- listen to a station broadcasting from some distant, future destination to get a sense for the events and culture of the area

Radio stations approached the 'net with some trepidation, since they knew it was an emerging medium with few restrictions and low barriers to entry. The trick, as most broadcasters saw it, was to use the Internet to complement their existing broadcasts. Typically, they simply wanted to extend their connection to their local listeners, providing additional advertising and promotional opportunities.

While conventional radio outlets were relatively quick to adopt Internet streaming in the late 1990s, broadcasters have charted a relatively cautious approach since then, as a recent Clear Channel decision demonstrates. Clear Channel, which now owns or manages some 1,200 radio stations nationwide, ceased corporate funding for radio station Web feeds in January 2003, forcing stations to use local budgets to continue streaming. With this decision, about three quarters of the Clear Channel stations ceased Webcasting altogether.*

While it seems clear that Web streaming may offer the potential to take radio in a whole new direction as convergence technologies evolve, broadcasters do not want the Internet to replace radio. Thus, you won't find many innovative new media plays or exciting new interactive radio formats coming out of the conventional broadcasting realm. This activity will be left largely to entrepreneurial creator-collaborators, including many who have already seized on the opportunity to create their own Internet "radio stations." These streaming media resources vary from a simple Web-accessible playlist of songs to an extensive effort that includes "personalities," giving some Internet radio stations a human identity – just like a *real* radio station.

---

* Randy Dotinga, "Radio Ditches Webcasts En Masse," *Wired News*, June 3, 2003 (http://www.wired.com/news/business/0,1367,59079,00.html).

The telecom act of 1996 eased restrictions on media ownership considerably. In the radio business Clear Channel expanded from an ownership base of 40 radio stations to owning or managing some 1,200. Clear Channel also operates many concert venues across the United States and a stable of television stations as well. Here in Syracuse, Clear Channel owns the top two rated radio stations in the Syracuse market, and four of the top six. It owns the top news/talk station (570 WSYR), the top-rated sports/talk station (Sportsradio 620), the top-rated country station (B 104.7), two of the most popular drive time FM music stations (Y94 FM and Hot-107.9), plus the local ABC television affiliate (WIXT 9). This sort of consolidation enables large media companies to control and coordinate conventional media outlets, strengthening this mass media, mass marketing platform.

In June 2003, the FCC moved to further relax rules on media ownership. These relaxed rules have now been challenged in court, and the congress is considering a resolution to repeal the new rules. See Frank Ahrens, "Senate Votes to Block FCC Media Rules," *Washington Post*, September 17, 2003 (http://www.washingtonpost.com/ac2/wp-dyn?pagename=article&node=&contentId=A21792-2003Sep16&notFound=true).

Many Internet radio feeds cater to unusual tastes, providing a global audience with specialized genres that you probably can't find on your local radio dial (Figure 3).

Figure 3. Internet "Radio Stations"

InternetradioHawaii.com	Hawaiian music
Bassdrive.com	Drum and bass radio
Beethoven.com	Classical
Boomer-radio.com	Baby boomer hits
Radioparadise.com	Eclectic intelligent rock
SomaFM.com	Groove salad
Digitally Imported	Electronic dance music genres
Smoothjazz.com	Smooth jazz from Monterey Bay
Hardradio.com	The heavy metal supersite
Philosomatika.com	100% psychedelic trance
MysticRadio.com	New age, chill, ragas, and more

Few Internet radio stations have been able to turn a significant profit, owing in large part to the audience fragmentation that is part and parcel of the 'net as a communications medium. Unicast streaming, which can incur significant costs for bandwidth, and royalty fees* add to the fiscal challenge of providing quality Internet radio feeds. While many Internet stations have a tough time making it, they continue to provide a unique and valuable entertainment experience for growing number of regular listeners.

* These fees were put in place in 2002 after a legal arbitration process that resulted from a protracted disagreement between Webcasters and the music industry. The parties have agreed on terms since then. From Roy Mark, "Webcaster, Music Industry Reach Royalty Deal," *Internet News*, April 4, 2003 (http://www.internetnews.com/bus-news/article.php/2175961).

Internet radio is especially interesting and potentially potent because it provides an incredibly diverse alternative to conventional radio and music television, allowing us to easily explore alternative music genres. My hope is that Internet radio will eventually take hold and, in combination with the vast and growing virtual library of Internet MP3 recordings and other digital music resources and technologies, usher in a bold new era of more accessible and more diverse music entertainment.

Inevitability

The extraordinary expectations for digital music will remain with us, no matter what the music industry does. There really is a new natural order. Many artists recognize this and have taken steps to embrace this reality.

Phish and other jam bands have offered Internet downloads of recent and archived concert recordings for some time now. The LivePhish.com service offers unrestricted, high-quality soundboard recordings of select Phish concerts in MP3 and CD-quality FLAC formats within 48 hours after the end of each show. Nugs.Net also offers concert recordings of several other bands in the jam-band genre such as the Dave Matthews Band, Leftover Salmon, String Cheese Incident, and Widespread Panic.

Rock veteran Peter Gabriel (who has his own record label and co-founded a legal music downloading service in Europe) and fellow-rocker Brian Eno have launched a music artist alliance, the Magnificent Union of Digitally Downloading Artists (MUDDA), to encourage and support artists who want to move to an artist-controlled 'net music download model (rather than, or in addition to, conventional CD marketing and distribution).* They're promoting the flexibility and fluidity of this new mechanism, which frees artists from the conventional CD distribution

* Associated Press Report, "Just Say No to Record Labels," *Wired News*, January 26, 2004 (http://www.wired.com/news/digiwood/0,1412,62050,00.html).

constraints imposed by the periodic release of multitrack music compilations.

Many creator-collaborators recognize that this new natural order will inevitably change how we create, distribute, share, and enjoy music. They call on the recording and broadcast industries to embrace this reality and help us to further extend the joy that music brings to all of us. Many others just go about the business of leveraging convergence technologies to listen to whatever they want, whenever and wherever they want to. A nice summary of the inevitability of a new digital music paradigm appeared in the conclusion of a white paper by Phil Leigh, who was featured in chapter 2.

> "The second coming of online digital music is as certain as tomorrow's sunrise. Every incremental gain in the number of broadband subscribers, or growth in the number of portable digital media players, or increase in the field population of CD rewriteable drives, takes everyone a step closer to the new dawn.
>
> Can it be avoided? Perhaps, but this will only happen if copy protection technologies are transparent to the legitimate users, and if they cannot be easily circumvented without detection, and if technological progress relating to computer devices such as rewriteable drives, portable MP3 players, and PC software can be halted or turned backward, and if millions of queued-up consumers discover that they really don't want broadband Internet access after all, and if owners of PCs will choose to stop using the 200 million or so CD burners that have already been shipped, and if buyers will choose not to be interested in using newer, smaller, and lower cost computer devices, and if, just once, the sun would rise in the West and set in the East. Admittedly the last event would be a great extraordinary sign, but in our analysis anyone who

believes in supernatural signs is going to spend the rest of their life being lost."

Excerpted from *The Second Coming of Online Music*
Phil Leigh, Vice President, Raymond James and Associates
March 19, 2002

While we are beginning to see where the new natural order of digital music may lead us in the realm of popular music, we are just beginning to see the tension building in the realm of video, which is the focus of our next chapter.

16. VIDEO: BIG CHALLENGES

Video is an extremely compelling medium that easily captures the attention and imagination of viewers and can be used effectively in a wide variety of business, entertainment, and educational contexts. While there is tremendous potential for video to enhance human experience during the second phase of digital convergence, video is also a big unknown when it comes to the convergence phenomenon because it is just that – BIG – when it comes to the volume of bits required to represent high-quality video. This creates a number of technical challenges for digital devices and networks. Video is also a big unknown because it is still too early to predict to what extent creator-collaborators will push the envelope when it comes to creating and sharing video and developing new video technology. Time, technology, and industry and social issues will all play into the equation that determines the breadth and depth of digital video deployment. One thing is certain: we will continue to see technology that provides ever-higher-quality video along with increased ease of use and flexibility – all at lower cost in terms of both time and money. Yes, the golden rule applies here as well.

Digital video is also huge because it will eventually serve as a marker of incredible progress. I believe that one day digital technologies will enable a more open media sharing platform – given the interactivity, media integration, universality, and flexibility that comprise the powerful underpinnings of the digital convergence phenomenon. Imagine a world of video that fits your needs and interests as well as your lifestyle and schedule:

- Instant access to a vast, high-quality entertainment video library that includes content from a wide range of independent video producers and artists as well as media conglomerates
- Easy, searchable access to video-on-demand content covering every imaginable news or educational topic that might help you to learn and grow

- The ability to quickly and easily slice, dice, and combine archival video clips and then share the results with individuals who share your interests
- Inexpensive and easy-to-use tools for creating and sharing quality video with colleagues, friends, and family
- Internet videoconferencing as a routine part of synchronous digital communications

When we someday find that this vision has been fully realized, I am sure that our lives and our society will have been fundamentally changed by the technologies of digital convergence.

I'm hoping we'll get there during the second phase – and I used to believe this was possible. Unfortunately, I'm now of a mind that it may not happen as soon as I'd hoped.

We will revisit the future of digital video at the end of this chapter. But first we need to explore the technologies and trends that are shaping the evolution of digital video as we move into phase 2. I'll break this discussion into four sections focusing in turn on broadcast television, desktop video media management, streaming video, and Internet videoconferencing. Of these four digital video application realms, television has been the dominant focus for consumers and for the entertainment industry for many years. But streaming video and PC-based video media management offer the greatest potential to fundamentally transform how we will experience digital video during phase 2.

Television

For several decades, conventional analog TV has offered what most of us perceive to be a very high-quality viewing experience – not because TV delivers high-resolution images (television is much lower resolution than the standard PC display), but because it reliably and continuously delivers full-color video to a large-screen device at an effective rate of 30 frames per second. Our perceptions are also influenced by the considerable resources and expertise that have been poured into lighting and audio and video production to ensure a level of video quality referred to in the industry as "high production values." As a baby boomer I grew up watching color TV, and I must admit I've been impressed (and hooked) for a long, long time.

Broadcast video is in the midst of a major move from analog to digital. Recording, editing, archival storage, and transmission are all moving rapidly to digital technologies. These technologies can provide higher-quality video, more programming, and increased interactivity and end-user control. Most visible to consumers are digital satellite TV, digital cable, and High Definition Television (HDTV).

Before we get into the evolution and capabilities of new digital video broadcast technologies, I want to highlight some of the shortcomings of conventional television. Conventional television has looked so good for so long, and become such a staple of our lives, that most people rarely ponder its foibles; but in light of the vast potential of digital video technology in the convergence era, conventional television is clearly an inherently flawed medium.

Why Conventional Analog Television Sucks

Television is a scheduled content delivery medium

Broadcasters and cable and satellite providers decide what content is transmitted and when. You have absolutely no control over this. If you want to watch *Judging Amy* and you aren't in the sofa in front of the TV when it's scheduled to be broadcast, you don't see it. Not unless you figure out how to program your VCR (or better yet, purchase a Personal Video Recorder, described below).

Television offers limited programming options

The objective of television programming is ratings, and broadcast programmers offer only that video content that is likely to appeal to a broad audience. Over the years the industry has identified a variety of genres that seem to work, and this has produced some quality video programming such as the annual Emmy Awards show, Olympics coverage, comedy programs like the *Cosby Show* and *The Simpsons*, and news programs like *60 Minutes* (your tastes may differ, of course). Unfortunately, there now seems to be a glut of relatively low-quality, lowest-common-denominator programming which consumes a considerable portion of a limited broadcast capacity – and frustrates many viewers who too often find that there just isn't all that much on TV. Take *Fortune* magazine's Peter Lewis, for example, who penned the following rant for a posting on his Weblog:

> "I suspect most [couch potatoes] will gobble up an expanded diet of 'reality' shows, vapid sitcoms, and shock-talk shows that feature semi-literate obese people in leopard-skin Spandex miniskirts talking about their favorite transsexual adventures with their daughter's boyfriends. That, at least, is what the broadcasters are providing now, because that, apparently, is what the audience is demanding."
>
> Peter Lewis, *Fortune.com*, August 2002*

Television enables only very limited interactivity

Increasingly, television programs refer to ancillary Web sites that broadcasters want you to visit. But with conventional analog TV there is no way to view and interact with the Web. Instead, you turn to your PC

* "Loony Tuners," Peter Lewis on Technology, *The Fortune Weblog*, Thursday, August 8, 2002 (http://www.fortune.com/fortune/ontech/0,15918,393868,00.html).

and Internet connection. Interaction via your television remote control is crude by comparison.

Television offers only video

You can't download an MP3 file, transmit an Excel spreadsheet, or access a Web page with text and images. Everything must be converted to one-way video (with synchronized audio).

No interactivity. No control over content or scheduling. No media other than video.

Now, consider alternative Internet-based convergence technologies – especially Web and streaming technologies. They can support video with full interactivity, integration with other media, and end-user control and customization. You can get what you want, when you want it, and wherever you want it. You can rather easily create your own video content and share it with your friends, colleagues, or anyone else who may share your special interests.

I hope you'll never look at your TV the same way again.

Digital Television (DTV): Better But Still Not Great

Digital television (DTV), the transmission of television content in compressed digital video format,* was pioneered by direct broadcast satellite (DBS) providers that delivered video at resolutions consistent with that of conventional television. Cable providers also subsequently launched standard-quality DTV services, and now an increasing number

* Within the television industry engineering community, DTV is a more specific term used to reference a group of interrelated standards. The ATSC (Advance Television System Committee) and the Digital Video Broadcasting Project (DVB) developed standard recommendations for digital television formats, capture, production, delivery, and so forth. These standards are used by technology manufacturers and have been adopted by government agencies such as the Federal Communications Commission.

of terrestrial broadcasters and cable and satellite providers are offering high-definition television (HDTV) broadcasts.

Digital television addresses some of the shortcomings of conventional analog television in three main areas: choice, quality, and interactivity. Let's look at each of these in a little more detail.

1. Choice

When direct-to-home (DTH) satellite television was first introduced to consumers, it required very large dish receivers that captured analog satellite signals from a number of geostationary satellites (which orbit the earth at a distance that ensures their orbits will match the earth's rotation). Eventually, direct broadcast satellite (DBS) services emerged to provide digital MPEG-2 compressed programming that was delivered to small-sized satellite receivers. These services offered hundreds of television and music channels.

Most cable providers have now upgraded their cable infrastructure (to a hybrid fiber optic–coaxial cable plant) to increase transmission capacity, and they also have begun to use digital MPEG-2 compression for digital video transmission. Thus, they too can offer hundreds of channels of television and music programs. With digital MPEG-2, satellite and cable providers can transmit multiple standard resolution programs over the radio frequency bandwidth of a single analog channel.

While hundreds of "channels" seems impressive to some, in my view this is still a very limited set of content sources. In the age of convergence, many more high-quality live and on-demand video sources could and should be possible.

2. Quality

High-definition television (HDTV) offers a clear step up in quality over conventional analog and digital television broadcasts. The most common high-def format has roughly twice as many horizontal lines of resolution, and the image format is based on the widescreen 16:9 aspect ratio of the movie business (the ratio between the width and height of the display – conventional television has an aspect ratio of 4:3, while movies are shot and displayed in theaters at 16:9). Terrestrial broadcasters are migrating to HDTV, and cable and satellite providers are delivering an increased amount of HDTV programming as well.

HDTV adoption moved at a snail's pace over the past several years amidst battles over standards, concerns over piracy, and pricing issues. HDTV adoption now appears to be turning the corner: as prices have finally declined somewhat, a critical mass of HDTV programs are available, and consumers are beginning to embrace this new format.* Expect this to accelerate over the next few years as prices continue to decline and HDTV quality wins over legions of consumers.

HDTV holds great hope for corporate mass media and mass marketing executives who seek to maintain control over consumer consciousness. However, the relative rigidity of the television broadcast model – especially in the realm of interactivity and constrained user control – could place the dominance of broadcast television at risk over the long haul.

3. Interactivity

There is incredible potential to use that large-screen television in your living room for a variety of compelling new applications. However, what has been introduced so far, by broadcasters and their satellite and cable partners, is just scratching the surface.

* The overall cost of HDTV capable televisions has declined fifty percent in the past couple of years. In 2002, there were 2.5 million digital TV sets sold, mostly HDTVs. That number is expected to grow to 4.3 million in 2003, and 30 million will be purchased in the next three years, according to the Consumer Electronics Association. From Steve Gelsi, "HDTV Reaches Mainstream," *CBS Marketwatch*, December 6, 2003 (http://cbs.marketwatch.com/news/story.asp?guid=%7BC4AFFD0C-7857-4867-B8B7-14F013AEF961%7D&siteid=google&dist=google).

With digital cable you can interact with a program guide, set parental controls, and order video on demand (VOD). VOD offerings include movies as well as popular programs such as *The Sopranos* from HBO. This is a fee-based program option that cable providers hope will generate significant additional revenue. The advertisements for Time Warner's iCONTROL VOD service, which is available here in Syracuse, touts the ability to fast-forward, rewind and (gasp) pause a video. While digital cable offers an improvement over analog broadcast or cable TV, it is still relatively crude when it comes to interactivity.

Satellite providers have been primarily focused on one-way video distribution for many years, with limited interactivity (ordering pay-TV services, for example) accomplished through a wireline return path, typically via a landline phone connection. However, they have begun to roll out interactive personal video recorders (PVRs are introduced in the next section), and more interactivity is on the horizon.

- EchoStar and SBC are working together to bundle DSL and satellite service that will ultimately provide for high-speed interactivity in a satellite TV set-top box*
- The DVB Project has developed a scheme for a satellite-based return channel, which would enable interactivity without the need for extra cabling of a wireline return.**

The European-based Digital Video Broadcasting Project (www.dvb.org) is a worldwide cooperative effort – by cable, satellite, and other video equipment and transmission providers – to develop industry standard digital television technologies that enable more flexible and functional digital television services. In the United States, CableLabs

* Colin C. Haley, "SBC Invests in EchoStar, Adds Dish TV Bundle," *Internet-News*, July 21, 2003 (http://www.internetnews.com/fina-news/article.php/2237731).

** DVB-RCS. For more see the DVB Web site (http://www.dvb.org) and Vendela Paxel, *DVB-RCS200: DVB With Return Channel Via Satellite* (http://www.dvb.org/documents/white-papers/dvbrcsarticle.pdf).

(www.cablelabs.org) is a not-for-profit enterprise – backed by the major North American cable companies and equipment providers – that is also dedicated to researching and developing new technologies for the industry. Both CableLabs, with its Open Cable and Open Cable Applications Platform (OCAP) initiatives, and the DVB Project, with its Multimedia Home Platform (MHP), are working to establish standard platforms for the development of interactive television services.

According to the CableLabs Web site, new interactive television services could include*:

- Electronic program guides (EPGs)
- Impulse pay-per-view (IPPV)
- Video-on-demand (VOD)
- Interactive sports, game shows
- Web access and features such as e-mail, chat, and instant messaging
- Interactive games
- Services such as shopping (television commerce, or "t-commerce"), home banking, and personal video recorders (PVR).

Cable providers and phone companies are also beginning to consider the use of Internet protocols for broadcast video delivery;** and IT in-

* From the CableLabs Web site, "Open Cable – Project Primer," (http://www.opencable.com/primer).

** Jan Colpaert, *Broadcast Video Delivery Over IP Networks (G1590A)*, a Scientific-Atlanta Whitepaper (http://www.scientificatlanta.com/products/customers/white-papers/1203-G1509A_LoRes1.pdf).

dustry heavyweights, including Microsoft,* offer solutions for video delivery over IP networks. However, the focus is typically on video delivery over a "managed" IP network (i.e., a private internet where network engineers can monitor and control network traffic). The primary benefits typically sought after by cable providers are cost reductions and a simplified architecture. Telecom providers considering a move into the video market are looking at IP as a cost-effective, flexible platform for new video services.

What I've described to this point is not exactly the kind of WOW! technology service that might be possible with an approach that emphasizes higher-bandwidth open Internet access to multimedia content, including high-quality streaming media.

While today's digital television providers envision a more functional interactive multimedia environment, the reality is that these efforts aren't pushing the limits on open interactivity and services, but are instead cautiously introducing new interactions that ensure high-quality and ease-of-use – with only a very limited portion of available transmission capacity and functionality designated for open Internet access. While this makes some sense for passive consumers, these efforts will disappoint a growing number of creator-collaborators who recognize the ability of convergence technologies to provide much more useful functions and interactions, and more flexible access to wide-ranging video content.

Some of the most interesting and potentially fruitful interactive television developments are occurring in a new generation of set-top boxes. Let's quickly review set-top capabilities and introduce some emerging set-top technologies that creator-collaborators may find particularly interesting and useful.

---

* Stephanie Olson, "Microsoft Forges Ahead With IP/TV Push," *News.Com*, October 9, 2003 (http://news.com.com/2100-1034_3-5089056.html?tag=st_rn).

The Set-Top Box

While first-generation satellite and digital cable set-top boxes basically decoded MPEG-2 video and provided an interactive program guide, newer set-tops offer much more.*

For example:

- Scientific-Atlanta, the cable set-top market leader, has equipped some Explorer series set-top models with high-speed Internet access (for Web surfing and email) along with USB and FireWire ports for connecting external devices such as video cameras.
- The Moxi Media Center, from Digeo (www.digeo.com), is an interactive television software platform that runs on Motorola set-tops. It includes game-playing features, a music jukebox function, and even an integrated Flash player. Both Charter Communications and Adelphia have deployed these advanced set-tops, with Charter having placed over a million boxes in subscriber homes.** Scientific-Atlanta is also working with Moxi Interactive developers to fit the Explorer MC home media center with Moxi interactive TV services.
- Other interactive television providers, such as OpenTV (opentv.com) and Liberate (liberate.com), also offer technology

* A move toward plug-and-play TV, where electronic cards can be plugged into digital TVs to provide cable or satellite service along with some additional functionality – without requiring a set-top box – is underway. While this would be convenient for consumers and cost-effective for cable and satellite providers, this will provide relatively limited capability in light of the advanced features appearing in new set-top boxes. From "What Plug-and-Play Means for TV," *BusinessWeek Online*, March 22, 2004 (http://www.businessweek.com/technology/content/mar2004/tc20040322_6039_tc119.htm).

** From a Digeo press release, "Digeo's Interactive Channel Service on OpenTV Platform Reaches More Than One Million Charter Cable Customers," February 2, 2004 (http://digeo.com/newsroom/index.jsp).

and services for implementing enhanced interactivity and t-commerce through set-top boxes.

Of course some of the most interesting and potentially useful new set-top capabilities – such as those that enable Internet communication and access to Internet resources – could take time away from conventional TV viewing and possibly disrupt advertising revenues. As a result, cable and satellite providers have been prone to move cautiously on advanced interactive TV services – unless they are pushed by the introduction of compelling alternative technologies.

When I started working on this book, an alternative set-top box had just emerged. Called a *personal video recorder* (PVR), or sometimes *digital video recorder* (DVR), this computer appliance enables you to record programming on its internal hard drive from regular conventional broadcast, cable, or satellite video sources. This seemingly simple innovation – pioneered by startup companies Replay Networks and TiVo* – turns out to be quite powerful.

A PVR periodically downloads a program guide for the local viewing area through a modem or broadband connection. It has a menu-driven user interface navigated via remote control. Through the menus you can easily pick programs that you want the PVR to record. You can pick an individual program, or you can configure the unit to record every airing of a particular show or series (on a TiVo this is referred to as a "season pass"). The PVR also records live television as you watch it, so you can pause live TV. When viewing prerecorded video you can also fast-forward through commercials.

I purchased a TiVo PVR from Best Buy as part of my "research" for this book. It's a Linux computer with an input for the cable feed and an output to the television. The user interface is designed for the television display monitor, and you navigate with a nifty handheld remote control.

---

* Malcolm Maclachlan, "Set-Top Box Makers Square Off at Showcase '99," *TechWeb News*, January 20, 1999 (http://www.techweb.com/wire/story/TWB19990120S0010).

I must say it's a great thing. I have it programmed to record the local news and the *CBS Evening News* every night, so that no matter when I get home, I can always catch the news. I initially set up season passes to *Law and Order, Austin City Limits, Nova, Frontline, The Shield* and (gulp) *The Osbournes*. I love being able to exert a little control over a device that has for so many years had too much control over me.

The PVR essentially changes the way you watch TV. You can easily pick and choose what you want to watch, and you get to decide when you want to watch it. While this sort of "time-shifting" is a good start toward on-demand content, you are still limited by the content available from local broadcasters, satellite providers, or the cable TV provider that serves your area.

After the considerable media attention that accompanied the rollout of TiVo and ReplayTV, several large technology providers snapped to attention: consumer electronics providers, such as Sony and Phillips, invested in PVRs; Dish Network and DirecTV began shipping PVR-equipped satellite receiver set-top boxes; and cable equipment providers such as Scientific-Atlanta and Motorola equipped their premium set-tops with PVR capability.

PVR technology hasn't gone unnoticed by the open source community. Open source PVR programs are available for desktop PCs; and the Telly interactive home media server, from Interact-TV (www.interact-tv.com), is an open source–based set-top that acts as a PVR, MP3 server, and photo display server. Interact-TV is actively recruiting software developers who may want to contribute to the evolution of this platform.

A PVR can indeed change the way you watch broadcast television, but this is just the beginning. Basic PVR capability will ultimately complement other emerging video sharing technologies, providing flexibility that goes well beyond the limited access and interactivity of broadcast television.

The latest PVRs can handle voluminous HDTV broadcasts. Scientific-Atlanta and hard drive maker Maxtor are developing PVRs that can record six programs simultaneously.* Some new PVRs are also DVD player/recorders that allow unlimited recording and archiving of broadcast television content. You can also program one of the newer Internet-connected PVRs remotely from a Web browser.

The evolution of the PVR, together with new technologies for interconnecting devices through a home network – through wired Ethernet, Wi-Fi, Bluetooth, and eventually ultra-wideband (UWB) connections – has stimulated technology providers in the consumer electronics, computer, and home networking sectors to introduce new multipurpose, network-based entertainment appliances.

For example:

- a DVD player that is also a PVR with a broadband connection to the Internet**
- a television that can display streaming content from a home-network–connected PC or from the Internet***
- a DVD player that can connect to a PC across a Wi-Fi network****

Broadcasters are already in a twirl because consumers with PVRs are fast-forwarding through commercials. They better get their feet on the ground sometime soon – much more extensive capabilities are in the works.

While the industry may pursue a fairly cautious and slow-moving approach to the introduction of new set-top capabilities, creator-

* Ed Frauenheim, "TiVo-like Devices to Get Booster Shot," *News.Com*, February 25, 2004 (http://news.com.com/2100-1041_3-5164465.html?tag=nefd_lede).

** Matthew Broersma and Richard Shim, "Sony DVR to Sport Broadband," *News.Com*, April 4, 2003 (http://news.com.com/2100-1041-995549.html).

*** Evan Hanson, Stefanie Olson, and Richard Shim, "Sony TV Would Grab Streams From the Net," *News.Com*, April 4, 2003 (http://news.com.com/2100-1025_3-995467.html?tag=st_rn).

**** John G. Spooner, "Gateway to Tout DVD Interaction," *News.Com*, June 27, 2003 (http://news.com.com/2100-1041-1021838.html?tag=nl).

collaborators are moving aggressively to explore very functional and powerful alternative platforms. For many creator-collaborators, a desktop PC as PVR and home media server has some real appeal, given PC flexibility and power. The ability for PCs to capture, create, edit, and share video across the 'net go well beyond what most set-tops provide.

Desktop Video Media Management

While increased video access and sharing through set-top and media server appliances is gaining some momentum, the PC has also staked its claim as control center for video access and sharing in the home. The PC, in combination with high-speed Internet access, will lead a surge toward a new natural order for video media, just as PCs, the Internet, and P2P technologies ushered in a new natural order for music.

Just for starters, the PC can become an incredibly flexible and functional PVR*:

- You can install a TV tuner card into your existing PC and purchase a PVR software program such as Show Shifter (www.showshifter.com) or SnapStream's Beyond TV3 (www.snapstream.com). With Beyond TV3 you can connect directly to an analog output on your cable or set-top box; with Show Shifter you can do that, or use a card that extracts MPEG-2 files directly from a satellite DBS or digital cable feed.
- nVIDIA sells a Personal Cinema bundle consisting of a graphics card and software that turns your PC into a PVR with the ability to capture, edit, and burn video.**

* For more on this, see Eric Hasseldahl, "Doing TV on the PC," *Forbes.com*, November 2, 2003 (http://www.forbes.com/2002/11/01/cx_ah_1101tentech.html).

** For more on this see the nVIDIA Web site (http://nvidia.com/page/personalcinema.html).

- Microsoft has come out with a Media Center version of its Windows XP operating system that supports video-enabled desktops. It includes software that allows your PC to serve as a PVR when connected to a broadcast video source. Most major PC manufacturers now sell desktop PCs bundled with Windows XP Media Center and the hardware required to connect any standard video source.
- The open source movement also has also developed PVR programs – MythTV, Freevo, and eBox – which run on inexpensive PC-based Linux boxes. For inexpensive and unconstrained access to recorded video programs, these are tough to beat.

High-quality video files – from DVDs or PVRs – can be stored on a PC hard disk, where they can be edited and integrated with other media files, and the results can be burned onto CDs or DVDs, or shared with family and friends via your home network or the Internet. Here is a quick rundown on some of these powerful PC-based capabilities.

Ripping: Sharing video got much easier when the source code for DVD encryption was hacked in 1999.* This, of course, led to widespread availability of DVD ripping software, which converts encrypted DVD video into PC hard disk files. DivX is a compressor that uses a combination of MPEG-4 and MP3** and is often used to compress ripped DVD titles (it is sometimes referred to as the MP3 of video). With DivX compression you can fit a feature-length movie on a CD. With a broadband connection, you can feasibly transmit DivX movies across the 'net, and with Napster-like peer-to-peer pro-

* Yoshiko Hara and Junko Yoshida, "Code Hack Delays DVD-Audio Launch," *Electronic Engineering Times*, December 3, 1999 (http://www.techweb.com/wire/story/TWB19991203S0005).

** Cynthia Flash, "DivX;-) May do to Movies What Napster Did to Music," *TechWeb News*, July 27, 2000 (http://www.techweb.com/wire/story/TWB20000727S0001).

grams, many creator-collaborators are now sharing DivX video files (including many popular movie titles).

Burning: Creator-collaborators can burn movies onto CDs using VideoCD (VCD) or Super VideoCD (SVCD) so they can be viewed on a DVD player. To do this you need a source movie (e.g., a DivX movie downloaded from the Internet), a CDR or CDRW drive on your PC, and software for creating and burning the SVCD or VDC.* The cost of PC-based DVD recorders is also declining rapidly, and it won't be long before this is standard equipment on mainstream PCs.

In-home transmission: Several vendors have created software and hardware solutions that enable you to stream video from your PC to other locations in your home over Ethernet or Wi-Fi networks. For example, SnapStream's Beyond TV3 includes a streaming video server function that works over wired or wireless networks, and Show Shifter enables you to convert MPEG-2 video to DivX or Windows Media format so it can be more easily streamed in the home or shared across the 'net.

Of course, creator-collaborators don't just view and share video; many also create video. With a digital video camera (and maybe some basic audio and lighting equipment), a PC with a nonlinear video editor, and a broadband Internet connection, individual creator-collaborators are able to rather easily create and disseminate compelling video content. This will eventually lead to a critical mass of Internet-accessible alternatives to the high-production-value, lowest-common-denominator video that is standard fare through conventional broadcast channels.

While dealing with the technical details of DVD ripping, DivX, and SVCD is well beyond what most passive consumers can handle, there is a growing creator-collaborator community that is focused on creating

* For more on ripping DVDs, burning SVCD CDs, etc., see the http://dvdrhelp.com.

and sharing video content and collaborating to develop technologies that enable us to manage video more easily and more effectively. PC video media management technologies will only get cheaper, more functional, and easier to learn and use over the course of the next few years. This is the beginning of Hollywood's nightmare.

Streaming Video

You may recall from the first book that I've been very high on the future of streaming media because the technology is relatively easy to deploy and it enables access to a global audience. However, many early efforts to capitalize on this new communications platform fell flat, and many streaming startups have now crashed and burned (mirroring the failure of many e-commerce dot.coms). So, before we look at what I think are some key streaming media success factors, let's quickly review of some of the factors that derailed so many first-generation streaming ventures.

Internet streaming video is not yet television quality

The quality of streaming video is still inferior to television at current broadband speeds (in the range of hundreds of thousands of bits per second).* Many early streaming video ventures, such as the failed Digital Entertainment Network (DEN), focused on television-like entertainment

* With a doubling or tripling of current broadband speeds, streaming video can compete very nicely with VHS quality video, and begin to push the envelope toward DVD quality. According to Real Networks, its RealVideo 10 codec can deliver HDTV quality at 75% less bitrate than current HDTV broadcasts, although current bandwidth constraints on the public Internet currently preclude the possibility of successful HDTV streaming. For more on this see the Real Networks Web site, *RealVideo 10 Technical Overview Version 1.0*, accessed on March 24, 2004 (http://docs.real.com/docs/rn/rv10/RV10_Tech_Overview.pdf).

experiences; many others attempted to simply reuse broadcast television video content (referred to in the industry as "repurposing"). Attempting to offer a television-like experience on the 'net doesn't play well to the strengths of this relatively new medium.

Viewers are not on the sofa

Access to Internet streaming video content has been almost exclusively via PCs. When you are using a PC, you are in an active "lean forward" mode. This is very different than sitting back and relaxing in your easy-chair or sofa to watch TV – a posture that is much more conducive for passive viewing of video content. Streaming media applications must take this into account* – at least until we have widely deployed PC-to-TV connections or set-tops that facilitate open access to Internet streaming content.

Scaling to serve many users is challenging

With unicast streaming, each and every person who watches a streaming video is sent a separate stream (this concept was introduced in the first book and is described here in chapter 12). For the streaming provider, this is problematic: serving streaming video to a large audience means significant costs for technology and Internet transmission capacity; and a suddenly popular stream can cause a traffic spike that may reduce the quality of the end-user experience. When compared with conventional broadcast technology, the cost/benefit when scaling to serve many viewers just isn't there yet.

* The Whitman School, in partnership with Sylvan Learning subsidiary Caliber, created a series of online streaming video courses that each consisted of a series of learning modules that were each in turn composed of very brief topic modules. Caliber had done significant research on user reaction to this format and had concluded that most people won't pay attention to a streaming video feed for more than seven minutes. Thus, our topic modules couldn't exceed seven minutes.

Given the realities described above, the streaming media industry has struggled to develop scalable business models. However, I believe that in time, streaming video will become a very important and widely used technology – there is a significant foundation of success to build upon.

Where Does Streaming Work Today?

I've highlighted some issues that have prevented streaming video from becoming established as a mainstream medium for the masses. This is where television excels – although it is still relatively expensive to produce and broadcast television content.

Streaming video, in contrast, is relatively inexpensive to produce and transmit when serving small audiences. Early, successful implementations have built on the unique strengths of the technology:

- Cost-effective for communities focused on a common interest, special interest groups, and organizations
- Global reach
- Easy integration with other media types and Web-based interactions
- Live or on-demand access
- An ability to provide fine-grained search access to indexed video media

Here are a few contexts where streaming media has been successfully deployed:

Corporate communications

Many companies use streaming video to reach out to their external constituents. For example, many stream corporate shareholder meetings to globally dispersed remote shareholders. Others offer on-demand promotional videos to customers who may be motivated to learn more about the company and its products. Many large corporations are willing to expend considerable resources to ensure that this is done well, and many successful streaming media companies have targeted this emerging market.

Distance learning

Distance learning, broadly defined, is an area ripe for the integration of streaming video. At the Whitman School of Management we've successfully used streaming to support executive learning and credit-based courses. We also launched a new Web site called *Business Connections* (whitman.syr.edu/connections) that is a home base for on-demand access to educational streaming videos as well as participation in live Webcasts featuring distinguished speakers, expert panels on current business topics, and other learning interactions within the school. Streaming enables us (and many other learning institutions) to cost-effectively reach out beyond the physical boundaries of our campus.

Corporate training

A combination of streaming media with integrated slide shows, online training documents, synchronous and asynchronous communication tools, and interactive assessments can effectively support a wide range of corporate training applications. For many companies – especially those that have a globally dispersed employee base – online training is one of the most cost-effective methods for employee training. Cameron Clarke, CEO of Vodium (a Washington, D.C., company that offers a suite of streaming media software and services), told me that corporate training is one of the largest applications of streaming media in the enterprise.

Customer support

Providing product information to customers via streaming video can be effective and efficient. This can include direct end-user access to training, product orientation, or troubleshooting videos.

Conference broadcasts

Streaming media enables conference and trade show organizers to add a remote, global audience for remote participants – typically highly motivated industry professionals and decision-makers. This enhances industry exposure for conference presenters and panelists, and provides another potentially valuable target audience for prospective advertisers.

These are just a few examples where streaming video has proven itself. The keys are the relatively low barriers to deploying this technology on a limited scale, and the ability to integrate streaming video with complementary media and interactions. As video quality improves, technical issues such as network QoS are resolved, and streaming is supported in a wider range of end-user devices (such as set-top boxes and hand-held devices*), we can expect to see streaming become a much more potent and pervasive communication tool.

Internet Videoconferencing

The public Internet infrastructure still lacks key technical attributes required for high-quality, reliable videoconferencing. As you can recall from earlier chapters, the public Internet does not ensure any end-to-end Quality of Service for real-time transmissions. Congestion can and does occur. That is one reason why a significant portion of business videoconferences still run over ISDN lines (circuit-switched digital phone lines). Internet videoconferencing is also problematic because

* While companies such as PacketVideo (http://packetvideo.com) have developed streaming solutions for handheld devices, I must admit that I'm still a bit skeptical when it comes to the overall potential utility of handheld video. Video in our homes and offices – on large-screen devices – is where the most significant developments will occur.

the end user has to set up a camera and configure his or her videoconferencing software.

The bottom line for Internet videoconferencing is that video quality remains an issue and tweaking the software and hardware to get things working properly can be complicated and time-consuming.* For many passive consumers, these hurdles are just too much to deal with.

In 2002 I watched the Windows XP rollout at the Hilton Times Square in New York City (via streaming media). Regis Philbin hosted the event, and it became clear that he was there, in part, to represent the millions of impatient and computer-phobic consumers who would presumably find Windows XP so much easier to learn and use than earlier versions of Windows. Regis was able to navigate several features of XP pretty well, but when Bill Gates sauntered out to Times Square to videoconference with Regis back in the conference center, sure enough, they couldn't get it to work. The support staff scrambled to cut over to a backup system and they eventually got the conference up and running. Unfortunately, this sort of glitch is pretty typical when it comes to Internet videoconferencing.

Beyond technical and reliability issues, one other significant barrier exists: social acceptance. Most people can understand the benefit of adding video to an electronic business conference, provided the quality is good. But many people are leery of a more general "videophone" capability for everyday use. People may want to use videoconferencing to see grandkids, or to flirt with a member of the opposite sex. But every time I bring this up in a class or presentation session, I find many in the audience (usually more women than men) who would not want to regularly be seen via videoconference.

Still, videoconferencing is seen as a promising technology by many. I've been waiting for many years for widespread adoption, and at this

---

* Mike Langberg, "Quality of Videoconferencing Isn't Clearing Up," *San Jose Mercury News*, March 1, 2004 (http://www.mercurynews.com/mld/mercurynews/business/8076246.htm).

point I've become somewhat of a skeptic, although there are some promising signs (e.g., university use of Internet-based videoconferencing has increased in recent years).

Perhaps creator-collaborators will begin to videoconference through the television with a 'net-connected PC, offering a much larger and perhaps more compelling video image. Or maybe the integration of videoconference capabilities in mainstream instant messaging solutions will build a significant user base. These are the sorts of developments that could turn the tide.

Now that we've reviewed several contexts where digital video is currently being used, let's reflect on where we're headed.

The Future of Digital Video

As I see it, the future of digital video – like so many other aspects of the emerging digital convergence phenomenon – will be shaped largely by the struggle between corporate control and the creator-collaborator culture of the Internet (Figure 1).

Figure 1

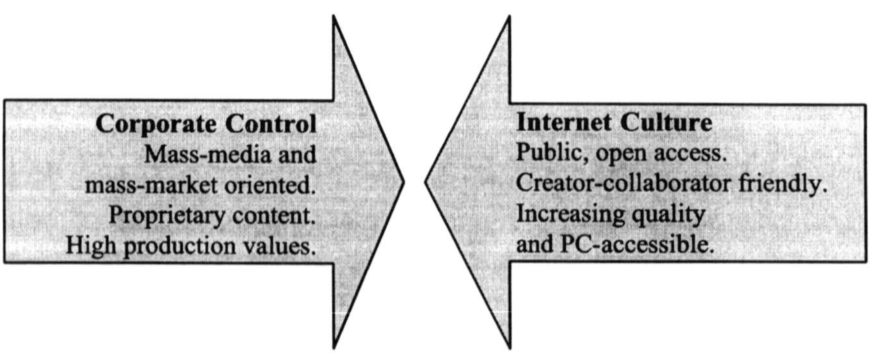

It often seems that corporate profit motive drives cable providers, broadcasters, and others in the direction of more control, not less. For example, the set-top box has the potential to provide access to high-quality content from the major studios and networks as well as open access to high-quality Internet streaming media content. Unfortunately, the cable providers that deploy set-tops are often unwilling to provide that open access. Instead they take a "walled garden" approach that provides fee-based access to only a select set of content resources. If conventional media companies and broadcasters are successful in retaining control over what we access through set-top boxes and our TVs, they can continue to direct mass audiences to lucrative, advertising-supported content properties.

Another effort at maintaining control over content, in the hopes of ensuring maximum profitability, is the development of digital media rights management (DRM) technology. While DRM technologies have become popular with many in the music recording industry – they may be even more coveted by Hollywood movie moguls.

As media conglomerates work vigorously to strengthen and extend control over digital video and other media, creator-collaborators continue to crank out new technologies for more easily accessing, sharing, and viewing quality video media. Of course, viewing video and sharing video is only part of the digital video equation. We also have new opportunities to create video content and to communicate more effectively with synchronous, real-time video interactions.

The creator-collaborator movement is empowered by the openness, flexibility, and interactivity of the Internet. Even as Internet bandwidth limitations continue to constitute a significant stumbling block when it comes to sharing high-quality video via the public 'net, creator-collaborators continue to work to reshape what video means to our work and our lives, and a steady stream of digital video content and technical advances remains – for now – "under the radar" of mainstream social and consumer consciousness. These advances will continue to provide increasingly accessible, functional, and useful alternatives to mainstream video media. And just as we have seen a new natural order for digital music, a new set of fundamental factors will ultimately create a new natural order for video, and drive a significant transformation. We can all look forward to a time when video becomes a much more useful, interesting, accessible, and pervasive communications and entertainment medium.

It's just a matter of time.

17. Mobile Internet

The hype machine has been turned up to full volume when it comes to emerging wireless technologies. Mobile Internet – staying connected on the go for voice, Web access, streaming media, instant messaging, or one of many other 'net-based interactions, services, or resources – is expected by many to be driving force behind the next wave of mass adoption of new digital technology. In light of the dramatic impact of the Internet/Web explosion of the 1990s, I remain skeptical when it comes to the overall impact of mobile technology on our lives, since I see this phenomenon focused largely on convenience rather than on facilitating whole new forms of human interaction that empower individuals and communities of interest. Still, I do expect to see continued advances in wireless technology, and I do believe that digital mobile connectivity will become increasingly important to many of us, although the grand vision of fully functional, anywhere-anytime mobile Internet access faces significant technical and nontechnical challenges.

Creator-collaborators in Western Europe, some Pacific Rim countries, and other places where factors have aligned more favorably may feel differently. The United States, which is used to being an information technology leader, follows many other countries when it comes to the penetration of wireless cellular technologies, one of the primary focal points for mobile Internet innovation over the course of the next few years. But I'm getting ahead of myself. We'll explore some of the geographic and cultural issues that impact perceptions and adoption of mobile technology around the world later on when we profile cellular use in the United States, Western Europe, and Japan.

Let's first survey promising mobile connectivity technologies and applications, which possess varying characteristics in terms of bandwidth, functionality, portability, and ease of use. We'll start by reviewing the most well known and widely used mobile devices.

Mobile Devices

The range of mobile digital devices commonly used for general purpose computing, communication, and information access is on the increase: from the tiniest digital cell phones with new computing and Web access capabilities to Sears-and-Roebuck-catalog sized laptop computers that have full keyboards, large LCD displays, and enough memory, storage, and CPU capacity to serve as capable multiuser servers. Here is a quick review of the most commonly used handheld devices.

Cell phones have been decreasing in size and increasing in capability for some time now. New phones feature cigarette-pack form factors and extended battery life. Still, the emphasis on decreased size and weight makes it challenging to deliver a device with significant computing power and a graphical, compelling display. Relatively limited text-based monochrome displays are still commonly used, and the full-color displays now featured by many cellular providers are still quite limited.

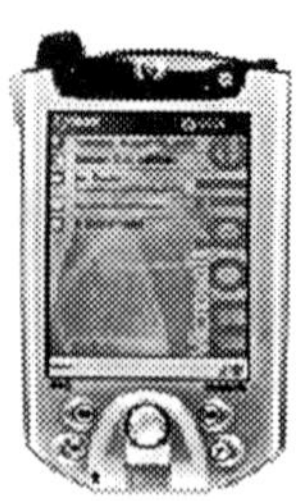

The Personal Digital Assistant (PDA) is rapidly evolving in power and functionality. Palm OS–based PDAs from Palm and Handspring, and Pocket PC-based devices from Sony, HP, and Dell have really caught on among business professionals who want to communicate and access information on the go. The Blackberry (from Research in Motion, RIM) is a popular alternative that offers always-on email access and a tiny keyboard (which many like better than the cumbersome handwriting recognition of other PDAs).

Notebook and laptop computers offer the complete functionality of the PC in a box that weighs just a few pounds. These devices are popular with many business road warriors, and they are also used by many others with active lifestyles and significant travel agendas, such as retirees and students (at the Whitman School we now recommend notebooks for all incoming freshmen and graduate students). With the increased availability of wireless networking (covered below) and other advances such as the power-saving, wireless and heavily promoted Intel Centrino technology, notebook-based mobile Internet access use is on the rise.

Beyond these basic mobile devices, the continuous introduction of new devices based on advances in miniaturization, functional convergence, and new capabilities is ongoing.

Here are a few examples:

Smartphones: Several converged devices combine voice with PDA capabilities. For example: the Handspring Treo is a Palm computer with built-in cell phone; Kyocera has produced a series of smartphones that combine a cell phone with a Palm PDA; the T-Mobile Pocket PC Phone Edition combines a cell phone with Pocket PC functionality; the Nokia Communicator offers a phone/PDA combo running on the Symbian operating system; the T-Mobile Sidekick is based on Danger's "hiptop" technology and provides always-on Internet access and supports Web surfing, email, and AOL instant messaging.

Tablet PCs: With a tablet PC you get the full functionality of a PC in a device that has a touch-sensitive screen that can swivel around to rest on the keyboard, a stylus, and handwriting recognition software – which together enable new types of mobile computing and interaction on the go.

Subnotebooks: While not as small as a pack of cigarettes, subnotebooks pack incredible punch in an incredibly small form factor, pushing full PC capability closer to the size and weight boundaries of the PDA. Sony, Panasonic, Toshiba, IBM, Dell and others offer subnotebook computers that include a 10" to 12" LCD screen and weigh as little as two pounds.

Wearable computers: When it comes to personal portability and access, wearable computers promise to raise the bar rather dramatically. While there are now only a few general-purpose wearable computers (such as the Fossil's Palm-based Wrist PDA, or the Poma wearable computer from Xybernaut technology, which features an innovative headmount display), many entrepreneurs and technology innovators are focusing on this exciting new frontier.

A student of mine did an interesting project report on wearable computers, which laid out the use of this technology in a variety of military and emergency service contexts, where there are some compelling applications today. He argued quite convincingly that wearable computers would end up becoming a common consumer accessory. Of course, when that will occur is debatable. While my gut tells me this will be later rather than sooner, my head tells me that many of the barriers to the adoption of wearables are quickly being overcome. The bottom line is that an increasing range of devices – with differing combinations of functionality, size, and weight – are providing ever-increasing mobile capabilities and options for creator-collaborators and consumers.

At this early stage in the evolution of mobile computing and Internet access no one knows exactly which combination of size and function will really catch on in a big way. Significant dollars are being invested – by individual entrepreneurs and venture capitalists on up to the largest major players including HP, Microsoft, Nokia, and Motorola – in the hopes of finding the platform that the masses will seize upon when mobile Internet really takes off. While many self-contained or "local" mobile applications are useful, most experts believe that future adoption will be driven by one or more killer applications that leverage network connectivity. So, let's now turn our attention to a review of emerging wireless network technologies.

Personal Area Networks (PANs)

PANs enable high-speed connections between devices that are in close proximity – from within the range of your immediate personal space on up to two or three dozen feet.

Infrared transmission is one fairly common PAN technology that you may have already encountered: an infrared network connection between a notebook computer and a printer can enable printing without wires; infrared connectivity can also enable people to beam information between two Palm Pilots. Infrared links are relatively slow, and they require line of sight – which renders this technology somewhat crude and cumbersome.

Bluetooth is a newer PAN technology developed by a consortium of computing and communications providers.* It offers much higher trans-

* The Bluetooth Special Interest Group (http://www.bluetooth.com).

mission rates using radio waves, which do not require line of sight. A Bluetooth connection can cover a distance of some thirty feet at a rate of 1 Mbps, and Bluetooth network connections are dynamically established through an electronic discovery mechanism. This technology can be used in a variety of potentially useful applications.

Here are a few quick examples:

Mobile commerce: Imagine a scenario where mass-marketed mobile Bluetooth-enabled devices would dynamically establish connections to receive advertising and sales information. Imagine walking through a mall and having your PDA or cell phone alerting you to special sales or digital coupon specials.* Or imagine being able to use your cell phone or PDA as an electronic wallet. You could electronically communicate with a vending machine to pay for and dispense a soda or candy bar – without ever touching the machine.

No more wires: Bluetooth is used in some "hands-free" phone headsets, and Bluetooth technology can eliminate the typical spaghetti-like mess of wires that connects your PC to cameras, printers, keyboards, digital video projectors, and other devices equipped with Bluetooth chips. Consumer electronics appliances, such as stereos, DVD players, and televisions, could also be interconnected with

* This vision of mobile commerce (or m-commerce) alarms me, and I suspect many of you are like me. To say that m-commerce zealots have their work cut out for them is an understatement – while there may be some option for creating a profile to weed out unwanted alerts, I can't imagine that this ultimately won't result in just more intrusive, unwanted spam. The International Telecommunications Union has also expressed concern and suggested that legislation may be necessary to protect consumers from potential abuses. For more information see the International Telecommunications Union, *Internet for a Mobile Generation: Executive Summary*, September 2002.

Bluetooth technology, and a Bluetooth-equipped PDA or smartphone could potentially serve as universal remote control.*

Internet access: A notebook computer or a PDA with a Bluetooth chip could connect to a Bluetooth-equipped cell phone (or some other Bluetooth-enabled network access point) to synchronize information and gain access to the Internet while on the go. The relatively low power requirements of Bluetooth chips would mean network access with minimal strain on notebook or PDA battery.

Bluetooth received tremendous hype in the trade and technology press when it first began to emerge as an option for personal network connectivity. However, early product deployments were slowed by problems with inconsistent and non-interoperable implementations and the emergence of increasingly popular Wi-Fi technology (described below). Bluetooth has gained some recent momentum as several new devices – PCs, hands-free cell phone kits, cell phones, printers, digital video cameras, even microwave ovens – are now being equipped with Bluetooth technology.

* The ZigBee Alliance –a consortium of semiconductor companies, software providers, system integrators, and original equipment manufacturers – is creating a similar standards-based wireless control technology that offers slower transmission speeds with low power consumption, so batteries last longer. ZigBee is initially targeting home control (e.g., lighting, heating, and cooling), industrial control, and industrial automation. From the ZigBee Web site (http://www.ZigBee.com).

At the February 2004 Intel Developer Forum Intel, Microsoft, HP, NEC, Samsung, and others announced their intent to cooperate in the development of Wireless USB technology. This is a promising alternative which is based on high-speed Ultra Wideband (UWB) technology, which was briefly introduced in chapter 13. From Edward F. Moltzen, "Consortium on Ultra Wide Band, Wireless USB Moves Forward," *Techweb.com*, February 18, 2004 (http://www.techweb.com/wire/story/TWB20040218S0015).

Wi-Fi

Wi-Fi, short for "wireless fidelity," is a popular term for high-speed wireless local area networks that are based on 802.11 IEEE* wireless Ethernet standard technologies and are certified by the Wi-Fi Alliance (www.wi-fi.org). Wi-Fi networks enable low-cost, high-speed connectivity; and while PAN networks provide network connections in a very short range, Wi-Fi networks provide connections in the range of hundreds of feet (this can be much less in some buildings, depending on the construction materials used). Carrying a notebook PC or PDA as you travel across the country on a business trip or out onto the deck at home can be useful. Having mobile access to a Wi-Fi network makes that PC or PDA *much* more useful.

While Wi-Fi technology is indeed very useful for local area wireless networking, a set of contiguous Wi-Fi networks can be coordinated to provide network connectivity for a much wider area. With widespread adoption of the Internet and increasing sales of Wi-Fi-equipped notebooks, Wi-Fi-based mobile Internet is becoming popular in a wide range of venues:

- Many corporate offices are implementing Wi-Fi networks to enable roaming employees to connect to the enterprise network and the Internet wherever and whenever they want.
- Wi-Fi has become popular in educational institutions where students with notebook computers can wander about with full network connectivity. (At Syracuse University we're currently implementing "AirOrange," a campus-wide Wi-Fi network).
- With the proliferation of inexpensive and easy-to-configure Wi-Fi access cards and hubs now available through all the major consumer electronics outlets, many consumers have imple-

* Institute for Electrical and Electronics Engineers (http://www.ieee.org).

mented wireless technology to enable device sharing and high-speed Internet access anywhere in the home.

- Some public spaces – including airports, libraries, and Starbucks coffee stores – are being fitted with wireless ethernet networks. These are often referred to as Wi-Fi "hotspots," and there are a growing number of fee-based commercial hotspots as well as free, community-based hotspots. Verizon has plans to convert phone booths in New York City to Wi-Fi hotspots. At Web-based hotspot directories, such as www.ezgoal.com/hotspots, you can find the hotspots in your area or some future travel destination. Wi-Fi hotspot providers such as Boingo Wireless and T-Mobile are supporting the growing trend toward open, public Wi-Fi access.*

While Wi-Fi technology is becoming very popular, it is still rapidly evolving, and some significant technical issues remain:

Multiple Wi-Fi standards: There are multiple wireless ethernet standards. IEEE 802.11b enables 11 Mbps shared communications, while IEEE 802.11a and 802.11g are more recent standards that both support a transmission rate of 54 Mbps. For large-scale installations, the choice of technology is far from straightforward.

Shared bandwidth: Wi-Fi is a shared bandwidth technology (every user shares the available bandwidth), which can mean local area

---

* As of January 24, 2004, Boingo had over 5,000 locations under contract and over 2,100 live worldwide. From the Boingo Web site (http://boingo.com/services.html).

T-Mobile has invested in Belair Networks, which has developed Wi-Fi mesh network technology that provides a cost effective solution for building out integrated metropolitan and regional Wi-Fi networks. From a press release on the Belair Networks Web site (http://belairnetworks.com/about_belair/press_releases_view.cfm?p_id=20).

congestion when users are concurrently transmitting or receiving too many bits for the network to handle. While this is not a problem in a typical home implementation, it can be problematic in a public implementation such as a university, airport, or metropolitan office park. Also, the extensive use of rich media applications can quickly swamp the available bandwidth of any shared Wi-Fi network. Careful planning and strategic deployment of access points and careful network monitoring can minimize the potential for congestion on Wi-Fi networks.

Unlicensed spectrum: In the U.S., the electromagnetic spectrum used for Wi-Fi networks is unlicensed – i.e., it is not licensed for any specific application or vendor by the Federal Communications Commission. This means that anyone can use this spectrum for any purpose. Thus, other network technologies that use the same unlicensed spectrum, such as Bluetooth-enabled devices and some cordless phones, can interfere with Wi-Fi communications. On the other hand, anyone can set up a Wi-Fi network without government approval – and this has enabled rapid, grassroots growth of Wi-Fi networks.

Security: Wi-Fi networks aren't always secure. Encryption and authentication mechanisms are sometimes implemented in corporate Wi-Fi networks, but many Wi-Fi networks set up by corporate departments or rogue users aren't secure. Many home Wi-Fi networks are also left open and accessible to anyone with a PC and a Wi-Fi card.

Coverage: Wi-Fi networks only work within relatively close proximity to an access point. This compares poorly with the broad geographic access areas of cellular networks, which are also evolving to support high-speed Internet access (wireless cellular services are described below).

One of the attractions of Wi-Fi is the potential cost savings when compared with the cost of a wired network. In the new Whitman School of Management building, we are exploring Wi-Fi networking in the classroom wing (i.e., a dense deployment or Wi-Fi access points, or maybe even a Wi-Fi access point on a cart for portable classroom use) as an alternative to wiring each and every seat. The cost savings could be significant, enabling us to move funds into other classroom technologies such as projectors, document cameras, and videoconferencing equipment.

Wi-Fi cost-effectiveness has also been behind a creator-collaborator-led phenomenon. A growing number of grassroots geeks have begun to configure Wi-Fi networks that share broadband connections among neighbors – in some cases organizing to provide coverage across entire neighborhoods and throughout significant portions of major metropolitan areas.* Broadband providers like Time Warner (which offers the Roadrunner broadband cable service) have fought back by identifying and threatening participants with fiscal sanctions and legal actions.** Creative, effective, and efficient solutions continue to bubble up from the trenches where creator-collaborators toil – exercising the corporate hair follicles of network giants in the process.

Cellular Networks

One of the most promising realms of high-speed mobile Internet access is emerging in the context of digital cellular networks.

The core concept of cellular networks is a series of contiguous and coordinated transmission towers. Each tower services the mobile users within a nearby radius, creating continuous geographic connectivity in a series of network cells, where each cell is served by its own transmission tower. Network connections are passed from cell tower to cell tower as

* For an example, see http://www.nycwireless.net.

** Peter Meyers, "Time Warner Cable Warns Wireless Users," *New York Times*, July 1, 2002.

callers travel about. Cellular providers tweak the arrangement of transmission towers to accommodate geography as well as traffic and usage patterns. In the United States, "nationwide" cellular providers T-Mobile, Cingular, Verizon, Sprint, AT&T, and Nextel operate networks that cover large geographic areas and enable seamless roaming with fairly reliable connectivity.*

The history of the cellular industry has seen successive generations of cellular technology (Figure 1).

Figure 1. Cellular Technology Generations**

1G	First Generation	Analog
2G	Second Generation	Low-bandwidth digital
2.5G	The current transition	Higher bandwidth (up to hundreds of kilobits/second)
3G	Third Generation	Broadband speeds (up to megabits/second)

First-generation cellular networks were analog, voice-only networks based on a technology called AMPS (Advanced Mobile Phone Service). In the 1990s cellular providers recognized that they could offer additional services with 2G digital transmission technology, and using digital

* Nationwide providers are those large cellular carriers that provide services in all regions of the United States. They do not operate cellular networks providing complete national geographic coverage, but each of the big six operates a network that covers at least 200 million subscribers, while the next largest provider covers less than sixty million people. From the Federal Communications Commission, *Annual Report and Analysis of Competitive Market Conditions with Respect to Commercial Mobile Services, Eighth Report*, June 26,2003.

** Initial implementations of 2.5G and 3G networks have not approached the upper limit on network performance. For more on this see Peter Rysavy, "Wireless Nirvana," *Network Computing*, October 21, 2002 (http://www.nwc.com/1322/1322f3.html).

audio compression technology they could also service more calls by using the same transmission capacity. The latter point was crucial, since it meant more revenue per cell tower. These second-generation networks provided data services as well as voice, but data services were limited because of bandwidth constraints.

In Europe a single digital standard was implemented for second-generation digital networks, called GSM (initially an acronym for Groupe Speciale Mobile, GSM now stands for Global System for Mobile communications). GSM networks support seamless roaming across the continent. European GSM networks also enable universal use of Short Message Service (SMS), a text messaging tool which has become very popular among European cell phone users.

In the U.S., second-generation digital cellular networks implemented by competing cellular providers used different transmission technologies (Figure 2). Some used Time Division Multiple Access (TDMA), some implemented Code Division Multiple Access (CDMA), while others implemented GSM.

Figure 2. U.S. Cell Providers: 2G Transmission Technology

AT&T Wireless	TDMA
Cingular	TDMA
Nextel	iDEN*
Sprint PCS	CDMA
Verizon	CDMA
T-Mobile	GSM

* Nextel was created using network spectrum originally designated for two-way radio communication (the FCC refers to this as Specialized Mobile Radio service, or SMR), and Nextel uses yet another transmission technology (called Integrated Digital Enhanced Network, or iDEN) developed by Motorola, which combines TDMA-based service with enhanced two-way radio communication. Nextel has been aggressively marketing the two-way "Direct Connect" radio capability as a special feature of its cell service for several years.

The diversity of technical transmission technologies in the United States created an inconsistent cellular network across the country. This effectively disabled data services that require interoperability, such as SMS, and it limited roaming. Unfortunately, U.S. cellular networks haven't benefited from the same sort of "coopetition" (a combination of cooperation and competition that seeks to build market share for all players) that propelled the standardized and now ubiquitous Internet infrastructure during the 1990s.

With the 2.5G and 3G networks now being implemented there are once again several alternative transmission technologies being deployed (Figure 3).

In the United States, cell phones have become a major factor in local calling and long distance markets. Two of the biggest draws have been (1) bundled long distance service and (2) flat fee plans with lots of "anytime" minutes. Several cell providers offer attractive nationwide long-distance plans, and Cricket (a subsidiary of Leap Wireless) is a relatively recent entrant that is hotly pursing local calling markets by offering unlimited local calls within its service areas. This has driven the average minutes of use per subscriber per month to around 500, much higher than in other parts of the world.*

I teach technologically savvy undergraduates and executives, and I've been asking since 2001 how many students have cell phones and what they use them for. The majority of students in both groups now own cell phones, but most continue to use them mostly for voice calls. Cell-based Internet access has been far from a smashing success here in the United States, although it is showing signs of catching on.

* From the Federal Communications Commission, *Annual Report and Analysis of Competitive Market Conditions with Respect to Commercial Mobile Services, Eighth Report*, June 26, 2003. Long distance, local, and pay phone segments of wireline telecommunications have all been losing business to wireless substitution.

Figure 3. U.S. Cell Providers:
2.5G and 3G Transmission Technologies*

	2.5G	*3G*
AT&T Wireless	GPRS	EDGE/WCDMA
Cingular	GRPS	EDGE/WCDMA
Nextel	WiDEN	Flash-OFDM**
Sprint PCS	1xRTT	1xEV-DV
Verizon	1xRTT	1xEV-DO
T-Mobile	GPRS	EDGE/WCDMA

In Western Europe and the Pacific Rim, 2G, 2.5G and 3G cellular 'net access and data services have been more successful. For example: in Japan, NTT DoCoMo introduced a very popular 2G data service called iMode, and KDDI has a new 3G data service that already has several

* The information in this table, with the exception of the Nextel data, is from the Federal Communications Commission, *Annual Report and Analysis of Competitive Market Conditions with Respect to Commercial Mobile Services, Eighth Report*, June 26, 2003.

** Wideband iDEN (WiDEN) technology will enable Nextel to upgrade its bandwidth without moving to an entirely new technology. For more information see:

Nextel news release, "Nextel to Deploy Higher Speed Data Technology, WiDEN Designed to Quadruple Packet Data Speeds," November, 14, 2003 (http://phx.corporate-ir.net/phoenix.zhtml?c=63347&p=irol-newsArticle&t=Regular&id=470345&).

Sue Marek, "Enhancements Extend iDEN Lifespan," *Wireless Week*, January 15, 2003 (http://www.wirelessweek.com/article/CA270761?spacedesc=Departments&st t=001).

Nextel is also reportedly considering moving to a much higher speed network technology that some consider to be "4G", based on Flarion's Orthogonal Frequency Division Multiplexing Technology (Flash-ODFM). From Andrew Orlowski, "Nextel Expands '4G' Trials," *The Register*, December 13, 2003 (http://www.theregister.co.uk/content/59/34505.html).

million subscribers; roughly half of South Korea's mobile phone subscribers used 2.5G services by the end of 2002; and in Europe, the high average revenue per user (roughly double the U.S. average) is attributed in large part to the use of data services (SMS leads the way while other Internet services are gaining in popularity). This extends a trend that saw 2G cell penetration lagging in the United States (Figure 4).*

Figure 4. 2002 Year-End Mobile Phone Penetration (Percent)

Italy	93
UK	85
Finland	85
Germany	72
South Korea	68
Australia	68
France	63
Japan	62
U.S.	49

There is a variety of reasons for this differential uptake of cellular technology in different regions of the world:

Coverage: The United States has a huge geographic area, and while areas where there is significant population have service, there are vast rural areas where there is no cellular coverage. In Europe and Japan there is less geographic area – and better overall coverage. Unfortunately, U.S. cell providers sometimes make it seem like there is na-

* Information on the relative success of Western European and Pacific Rim data services along with the table on penetration rates are from the Federal Communications Commission, *Annual Report and Analysis of Competitive Market Conditions With Respect to Commercial Mobile Services Eighth Report*, June 26, 2003.

tionwide coverage with "nationwide" roaming deals and misleading commercials ("can you hear me now?").

Pricing: In Western European countries, the calling party pays. You can get a prepaid phone in Europe, and once you get the phone you can take incoming calls without charge. In the United States you pay not only for outgoing calls, but for incoming calls as well. Your monthly payment typically provides for unlimited night and weekend usage along with some additional "anytime" minutes, and you are charged significant fees for calls outside of your service plan. Fortunately, recent U.S. cellular competition has been pushing providers to lower pricing and provide some extended "anytime" minutes and all-you-can-eat plans that have begun to eat away at price as a barrier to adoption.

SMS Service: In Western Europe, Short Message Service (SMS) is inexpensive and it is interoperable across all carriers. In the United States, SMS service has not been universally and seamlessly implemented across all carriers, so it has never emerged as an important driving force for adoption.

Culture: In some parts of the world cell phones quickly became a status symbol of the young and young at heart. In Japan, some of the most popular digital downloads are cartoons and games. A professor I work with told me that when he visited South Korea he was amazed at the widespread adoption of miniature cell phones among Korean teens, who wore them around their necks like high-tech necklaces. U.S. teen culture is not yet so tied to the cell phone. In many U.S. cities and suburbs instant messaging is the "in" thing, and access to cable-modem- or DSL-based broadband is seen as a status symbol. U.S. PC and PC-based Internet access penetration rates are greater than in most other parts of the world.

Opportunity: In Japan, there is extensive use of mass transit. This provides many Japanese with a window of opportunity every workday to access information and communicate via cell network. Contrast that with the automobile-oriented suburban sprawl of the U.S., where daily commutes provide an opportunity for phone calls, but no opportunity for mobile Internet access.

My Own Experience

A few summers ago I went out and bought a Kyocera smartphone for $500 (with a one year cellular contract) to learn more about PDAs and cell phones and the coming wave of cellular data services. I wanted a chance to check out what were then state-of-the-art 2G data services and mobile device interactivity. The phone is both a Palm PDA and a cell phone. It enables me to keep my calendar, maintain my contact lists, use the AvantGo Web clipping service to keep up with news on the go, access my email, and surf the Web.

I chose Verizon as my cellular provider since they have better network coverage in my area of central and northern New York (although I still don't have to travel very far to lose service).

The smartphone helps keep me organized and in communication with the important people from my professional and personal life. I do like being able to look up phone numbers, and for a while I would regularly synchronize the phone with my desktop PC so that I would have on-the-go access to my schedule. However, I find that I rarely need access to my schedule when I'm not at work (where it's available right on my desktop PC), and I haven't synched my phone with my desktop in months. I've also encountered several problems with the phone and its services:

- *Downloads are s-l-o-w.* The limited bandwidth of Verizon's 2G network couldn't provide a useful service so far as I was concerned. I get a lot of email, and downloading new message headers took too long, so I canceled this add-on service. With 2.5G and 3G networks, which provide "always-on" connectivity, it's possible to have email pushed to your handheld device as it arrives, so that you no longer need to download email en-masse. Still, it may take quite a while before these upgraded networks provide adequate bandwidth for really effective Internet access: 2.5G networks, which originally promised hundreds of kilobits per second transmission capacity, are actually offering speeds similar to narrowband dialup modem rates; 3G networks, which have promised to offer transmission rates in the megabits per second range, will initially deliver only the hundreds of kilobits per second once promised for 2.5 G networks.*

- *It's hard to read.* The screen size is just too small and the legibility of the display is poor at best. Here the tradeoff is between screen readability, size of the unit, and power consumption and battery life. The newer screens are brighter and in full color, but they are still very small.
- *Data input is cumbersome.* I learned the standard Palm handwriting recognition language, called Graffiti. Unfortunately, I find I sometimes forget characters and I often need to re-enter characters several times to have them properly identified. Some technology providers suggest that voice recognition will overcome this deficiency, but I don't buy it. My Kyocera phone has that too – I trained it once to call my son and then found that when I *really* needed to use it (while driving, with considerable background noise, and at a considerably higher stress level), it just didn't work. I ended up screaming at the phone like a madman. I think a keypad that enables thumb-typing, like on a Blackberry or a T-Mobile Sidekick, is probably a better solution.

* Peter Rysavy, "Wireless Nirvana," *Network Computing*, October 21, 2002 (http://www.nwc.com/1322/1322f3.html).

- *The flip isn't durable.* I bought a little belt carrier, but it is big enough (it's about the size of a pack of cigars) that I just couldn't see myself sporting it on my belt loops. So I usually carried it in my pants pocket. One day the dialpad, which is on the outside of the flip, just quit working. I got a replacement since it was on warranty, and the second unit eventually suffered the same fate. I'm on my third phone and no longer carry it in my pants. Successful devices will have to deal more effectively with modest abuse.
- *It is expensive to use.* One month my bill was $193.00! Not only is it expensive, but figuring out my bill, which is often over ten pages, is like decoding a secret language. I once called a customer service rep to go over all the codes and charges. I got a basic understanding and learned about some charges I was totally unaware of (e.g., foreign towers) in about 45 minutes. I like my Roadrunner bill much better: I pay a fixed monthly fee, and it doesn't matter *how many* bits I send or receive and *where* on the planet the bits go to or come from!

Third-Generation (3G) Cellular and the Search for a Killer App

2G network bandwidth is limited, and the large, nationwide U.S cellular providers (as well as large European and Pacific Rim cellular providers such as Deutsche Telekom, British Telecom, Vodaphone, Japan Telecom, and NTT DoCoMo) are all implementing 2.5G or 3G cellular networks that offer much more bandwidth than is needed for basic voice service. So, what will that bandwidth be used for? There is concern among all cell providers that consumers may not bite when it comes to next-generation data services, especially here in the United States. Access to stock prices, weather reports, sports scores, and canned video programming just won't cut it. It seems that 2.5G and 3G-based services need some killer applications to drive widespread adoption.

I've had the 3G "killer app" discussion with every undergraduate and executive class I've taught for the last two years, and there has never been any real consensus on any of the applications promoted by cellular carriers. This is where 2.5 and 3G network providers need to take some cues from the success of the open, public Internet.

Figure 5. Mobile Service Availability by Device*

	Mobile Phone	*PDA*	*Laptop*
Limited Web browsing	X		
Open Web browsing		X	X

The Internet leverages the creativity of not only well-funded corporate content and service providers, but that of creator-collaborators as well. To be successful, 3G cellular services must follow the lead of the wired Internet to deliver a similar empowering experience for creator-collaborators and consumers. Handheld devices need to be like a PC Web browser – open to any and all comers. Music sharing, telephony, information sharing, transactional capabilities, and rich media experiences of all types need to be accessible from a rapidly growing base of sources. Otherwise, why would someone pay extra for 3G service on a handheld gadget? While notebook and PDA providers understand the

* "With mobile telephone handsets, web browsing is generally limited to the web sites offered by the content providers with whom a carrier has a content agreement. Therefore, most mobile telephone carriers allow web users to access a variety of popular web sites and applications on their mobile handsets but do not allow access to the entire web." Also, see Appendix E, Table 3. From the Federal Communications Commission, *Annual Report and Analysis of Competitive Market Conditions With Respect to Commercial Mobile Services, Eight Report*, June 26, 2003.

open access mantra, phone carriers have continued to push bundled Web and Internet services that are under their control.

Virgin Mobile, a joint venture of Sprint PCS and the Virgin Group, is an example of a provider that offers branded content and interactions to the MTV generation, including these exclusives: the "hottest tracks" for ringtones, voting on popular music and pop culture issues, and celebrity voicemail recordings.* These efforts to bundle specific content and limited services into cell handsets and service plans are designed to secure market share, but in the process they undermine the overall usefulness of a network-connected handset, which has the potential to be an important open Internet access platform.

As another example, consider the recent rollout of wireless video services by cell providers.** Sprint has introduced a service called MobiTV that provides unlimited viewing access to fourteen cable channels including MSNBC, The Discovery Channel, CNBC, and The Learning Channel. To use the service you must subscribe to the PCS Vision plan and then pay an additional $9.99 per month. Unfortunately, your choices are limited to the fourteen channels MobiTV provides. AT&T, Sprint, Verizon Wireless, and Nextel also provide access to something called 1KTV, which offers access to a library of short video clips from the Weather Channel, Fox Sports, and others – for $2.95 per month; and Real Networks provides a library of audio and video content developed by partners ABC News, CBS Marketwatch, The Sporting News Radio, and others – for an additional $4.95 per month.

This introduction of fee-based access to a very limited content menu contrasts rather dramatically with the introduction of Internet streaming services. When streaming first emerged on the open public 'net, streaming players were introduced with a variety of feeds to demonstrate the

---

* From the Virgin Mobile USA Web site (http://www.virginmobileusa.com).

** Mike Langberg, "Oh No! Live TV on Cell Phones Seems Silly but It's Here and It's Going to Get Better," *San Jose Mercury News*, February 23, 2004 (http://www.mercurynews.com/mld/mercurynews/business/8019184.htm).

concept. Encoders for creating streaming content were also made freely available, and original streaming technology pioneers – such as Xing, Real Networks, and VDOLive! – encouraged widespread development of Internet-accessible streaming content. Unfortunately, this model hasn't been fully embraced when it comes to new rich media services based on 2.5 and 3G networks.

Unfortunately, in the cellular network business, "strategic" vendor-specific applications – including limited Web browsing with featured content, exclusive catalogs of downloadable games, vendor-specific "push-to-talk" walkie-talkie features,* and phone-cameras that enable users to instantly send pictures to other users, but only on the same service provider network – point toward a more closed, proprietary client interface and experience. The notion of capturing market share by offering closed, proprietary services has taken hold in the cellular world.

It is obvious that all mobile Internet providers, including cell providers, must open the client interface for professional developers. But they also should facilitate and promote authoring and hosting for the masses. I'm convinced that with an open, inexpensive, and user-friendly mechanism for creating mobile content and interactions that are targeted to a wide range of handheld devices, we would see an explosion of handheld mobile Internet access – similar to the Web explosion we saw in the 1990s.

It is also my hope that someday the creator-collaborator–friendly IT industry approach for developing and marketing platforms – focusing on physical and intermediate layers, standardizing intermediate application

* Nextel, which uses spectrum originally intended for radio dispatch (Specialized Mobile Radio, or SMR) has been offering push-to-talk for some time. Verizon Wireless and Sprint PCS have also introduced push-to-talk features, but they only work between phone handsets on a single cell provider network. Other providers are also planning push-to-talk features, and while a push-to-talk standard has been under development (PoC, or push-to-talk over cellular) support for interoperability among nationwide network providers is far from certain since this feature differentiates cellular service provider offerings.

and transmission technologies, and then opening the floodgates at the application layers – takes hold across the converging telecom, cable, and wireless sectors.

The Satellite Wildcard

Satellite mobile network technology hasn't yet emerged as a mainstream option. However, this slowly developing technology is still very promising, largely because it has the potential to provide a global roaming capability. Thus, it earns a place as a wildcard mobile Internet technology that could become important down the road.

There are some positive signs:

> *Satellite services may soon be profitable.* Iridium, the $5 billion Motorola-backed venture, was one of the first to implement a global satellite phone service, but went bankrupt in under a year. The network still exists, and the new privately owned Iridium Satellite is moving rapidly toward profitability (the bankrupt venture was bought by investors for $25 million in 2002; the new owners indicated that they expected to be in the black in 2003).*

> *Costs for satellite access are coming down.* Globalstar has introduced a $34.95 phone service which offers 30 "free" minutes, with additional minutes at $.99 per minute.**

> *Mobile satellite providers are actively pursuing high-speed Internet access.* London-based Inmarsat has launched its Regional BGAN

* Christine Chen, "Iridium: From Punch Line to Profit," *Fortune*, September 2, 2002.

** From the Globalstar Web site (http://www.globalstarusa.com/pricing/home_rates.shtml).

service, which provides Internet access at twice the speed of 2.5G cellular networks within a 99-country satellite footprint.*

Wi-Fi Has the Early Edge

I have described a range of mobile devices and focused considerable attention on Bluetooth, 3G cellular, and Wi-Fi mobile network technologies. While it is too early to say which mobile network technology or technologies will dominate in phase 2, I give Wi-Fi the edge as the most promising mobile network technology for creator-collaborators.**

- Bluetooth is still fairly nonexistent (I have yet to actually use a working Bluetooth connection; hopefully I'll get the chance sometime soon), and the promise of Bluetooth lies largely in the convenience of networking local devices together.
- 3G networks, while offering high-speed transmission over a considerable geographic area, have been slow to offer the fully functional, open Internet access that Wi-Fi networks provide. The small-sized devices with tiny displays typically targeted for 3G networks also limit the usefulness of this potentially potent network service. And add-on pricing for cell-based Internet access is another barrier to widespread adoption in the United States. Add to that the closed-access approach of many new advanced services now being rolled out, such as video, and you can begin to see the how the utility of 3G may pale in comparison to open, high-speed Wi-Fi access.

---

* From the Regional BGAN Web site (http://regionalbgan.inmarsat.com/page_1.htm).

** For another take on why Wi-Fi has an early edge, see Clay Shirky, "Why Wi-Fi Will Clobber 3G: Lousy and Cheap Beats Good and Expensive," *Internet Week*, April 2, 2003 (http://www.internetweek.com/story/showArticle.jhtml?articleID=8600075).

- Wi-Fi networks provide notebook computers (which are routinely being equipped with Wi-Fi cards) and an increasing number of PDAs with open, mobile broadband Internet connections – making these devices much more productive and useful for creator-collaborators. The rapid, grassroots growth of Wi-Fi networks has underscored the utility of this flexible, inexpensive network technology. Wi-Fi offers a wide-open opportunity for innovative new forms of mobile interaction, collaboration, and information sharing to emerge. This is the bread and butter of the creator-collaborator movement.

18. Media Integration, Interactivity, Accessibility, and the Desktop PC

As phase 2 proceeds, the most important digital convergence capabilities will continue to surface in these three areas:

1. *Media integration*: text, image, audio, video, animation, and other digital media can be combined in any manner and delivered to any device at any time.

2. *Interactivity*: no convergence technology or interface need be constrained to any sort of limited, passive use. Interactivity complements media integration enabling customization, communication, and enhanced functionality.

3. *Accessibility*: the Internet reaches across the face of the planet, providing high-speed, global access to media and interactions from an increasing range of fixed and mobile devices. Convergence technologies are also accessible in the sense that they are readily available, inexpensive, and increasingly easy to use.

With virtually unlimited options for combining capabilities across these areas, the digital convergence phenomenon enables us to move far beyond anything that has come before – in terms of how we communicate, interact, and conduct commerce. Convergence technologies are also rapidly evolving – with a constantly changing set of platforms, functions, form factors, and tools. While I have suggested some things that I think might happen in the future – or that I hope will happen – there are some exciting and totally unexpected developments waiting in the wings.

The pace of technological change and the ever-increasing and ever-changing choice of technologies are somewhat daunting – and really just a bit too much for passive consumers. But you are different. As a creator-collaborator, or aspiring creator-collaborator, you have what it takes to actively explore convergence technologies that can take your life in exciting new directions.

Powerful Capabilities for Creator-Collaborators

In chapter 16 I conveyed what I believe are the most significant shortcomings – including inadequate media integration, limited end-user control, and minimal interactivity – of conventional television. Now consider the list of tools and capabilities shown in Figure 1. With the right motivation, some time, and maybe a little help, you can successfully pursue any of these – and share the results of your efforts in an integrated, interactive fashion via the Internet and the Web. Convergence is not just another platform for mass media communication and the building of mass markets.

I found an interesting interview on the *Wired News* site with David Byrne, former member of the rock band Talking Heads, who has begun to use PowerPoint – leveraging its full-screen display and its ability for animating and integrating visual elements, text, and music – as a creative, artistic tool. In the interview he offers some useful comments on the notion that empowering capabilities exist in some very inexpensive tools – like PowerPoint – that are very easy to learn and use.

> "I'd like for people to feel less timid about taking things into their own hands. This is exactly what was intended by the people who developed this program. They hoped that this tool would allow people to bypass the middleman, to communicate without having to work through a gauntlet of graphic designers or AV professionals. Do it yourself. After all, I learned how to do it in only a couple of hours."*

* Xeni Jardin, "Turning Heads With PowerPoint," *Wired News*, December 9, 2003 (http://www.wired.com/news/culture/0,1284,61485,00.html?tw=wn_story_page_prev2).

I hope you take these words to heart. There really is so much you can accomplish with these new tools – even without the benefit of professional training.

Figure 1. Convergence Tools and Capabilities: A Sampling

Real Media Helix Producer	Develop your own streaming media content
Flash	Create animations that are instructive, expressive, or entertaining
FrontPage	Create a well-organized, interactive Web site
SHOUTcast	Set up your own Internet radio station
PayPal	Process credit card payments for your online merchandise
Photoshop Elements	Share your best digital photos in a Web-based photo album
ShowShifter	Add Personal Video Recorder capabilities to your PC
Windows Messenger	Videoconference with friends or colleagues
WebEx	Deliver a real-time, Web-based seminar, training session, or sales pitch
Windows Media Player	Build and manage your own digital music or video library
Radio Userland	Set up an interactive Weblog to share your insights with the world
Adobe Premiere	Shoot and edit a video on your desktop computer

While there are many opportunities for untrained creator-collaborators to use very capable new convergence tools, there is also an army of media and Web professionals who are constantly pushing the envelope to create slicker and more sizzling interfaces and resources. I refer to this as "slickware."

Don't Let Slickware Deter You

Corporate media and Web providers embrace slickware as a means of differentiating themselves from the competition – and from the more plebian content and interactions developed by creator-collaborators.

- In the realm of conventional media, the most obvious examples are in video, where costly special effects technologies have enabled video pros to create not-easily-replicable video content. This runs the gamut from entirely fabricated video elements to glitzy transitions. You can't even watch a football game on TV without hearing a low swoosh sound with a sophisticated, sometimes comical animation every time a new graphical display appears or there is a transition to an instant replay.
- The phase 2 Web features tons of glitzy corporate sites that are created and maintained by full-time Web-design professionals. This level of technical sophistication – which includes extensive use of Flash animation, Javascript-based navigation and interactive features, cascading style sheets, and so on – goes well beyond what many creator-collaborators can tackle, given the significant time and expertise required.

Don't be deterred by your inability to match the slickness of professional media or Web resources. Like most creator-collaborators, you will probably be connecting with a niche group of colleagues, friends, coworkers, clients, or customers – and your focus should be on communicating and creating resources to serve your mutual interests. The desktop computer offers all you need to provide simple, clean, and functional communications, media, and Web interactions – and if you focus on providing just that, you can surely succeed.

The Swiss Army Bazooka of the Modern Era

> "Well, the device I really wanted to end on is by far the smartest and most important device in any home and that is, of course, the PC."
>
> Bill Gates, in his remarks at the International 2003 Consumer Electronics Show*

In chapter 16 we reviewed set-top box capabilities and the PVR, and I introduced some new digital-media server appliances. Game consoles are also becoming very capable – with broadband connectivity options and software that enable general-purpose Internet access capabilities (they are increasingly popular in the home, especially among young males).** In chapter 17 we also explored a variety of new handheld mobile devices.

All of these convergence devices will be increasingly important in the coming years, and this has caused some to ponder the end of PC dominance. While some passive consumers may prefer the ease-of-use of set-tops, the entertainment value of game consoles, or the mobility of handheld devices, the desktop PC will remain the primary platform for most creator-collaborators working to exploit new convergence capabilities.

The desktop PC is built from the ground up to connect to all sorts of other devices and networks. What's more, a wide variety of PC software is available that leverages the interconnectivity and capabilities of these other devices to provide useful new functionality. Thus, the PC will

---

* From Bill Gates' Web site, "Remarks by Bill Gates," 2003 International Consumer Electronics Show, Las Vegas, Nevada, January 8, 2003 (http://www.microsoft.com/billgates/speeches/2003/01-08ces.asp).

** John Borland, "Future: A New Prime Time," *News.Com*, December 11, 2003 (http://news.com.com/2009-1043_3-5116866.html).

emerge in phase 2 as a powerful and flexible control center for creating, managing, and integrating digital media and interactions.

People like you and me will utilize increasingly powerful desktop PCs – in concert with a variety of other convergence devices – to explore and exploit powerful digital capabilities and resources. I honestly believe that the results will be significant. And I hope that when the dust settles on digital convergence phase 2, my efforts to inform, guide, and energize creator-collaborators will have made a real difference.

INDEX

About the Author

Andy Covell is the author of *Digital Convergence: How the Merging of Computers, Communications and Multimedia is Transforming Our Lives (Aegis Publishing, 1999)* and numerous articles for *Network Computing* magazine and other technology trade publications.

Covell has been an Information Technology professional at Syracuse University for over two decades. He joined the Martin J. Whitman School of Management in 1992 where he is currently the *director of information technology*. He has been involved in Syracuse University institutional technology planning, served on several University technology task forces, and been a technology consultant for other Syracuse University schools and colleges.

At the Whitman School, Covell has developed executive seminars and courses for undergraduates and independent-study MBA students on digital convergence, digital commerce, and entrepreneurship.

In the 1990s he founded University Technology Associates (UTA) which offered technology consulting services to Central New York businesses and educational institutions.